2-24-01

ODOR of W~~

Elaine:

Thank you & all your family — great to meet you —

As ever

Andy

Ambrose

ODOR *of* WAR

SERGEANT ANDY GIAMBRONI

*I dedicate this book to my family
and to the many war veterans
who gave so much!*

Contents

PREFACE   9

THE HEART OF THE INFANTRY SOLDIER   11

1   TRAINING IN CALIFORNIA   19

2   SHIPPING OUT TO ENGLAND   23

3   CROSSING THE CHANNEL   27

4   COMPANY B—NANCY, FRANCE   29

5   METZ—THE LORRAINE COUNTRY   33

6   CHRISTMAS IN BELGIUM   41

7   MAGERET: FIRST ENGAGEMENT   47

8   SECOND ENGAGEMENT AT MAGERET   53

9   ARLONCOURT   61

10 THE COLD ARDENNES   67

11 THE OUR RIVER   73

12   LUNENBACH   79

13   MILLION-DOLLAR WOUND   91

14   COLONEL JIM MONCRIEF   97
       BUCHENWALD
       A SURVIVOR OF BUCHENWALD

15   RETURNING TO THE FRONT   107

16   THE OCCUPATIONAL ARMY   113

17   GOING HOME   119

18   THE LAST LEG HOME IN THE US   125

       CLOSE COMBAT   131

       BIOGRAPHICAL SKETCH   134

       CRITIQUE   137

       TAPS   141

       ACKNOWLEDGMENTS   145

I am writing this book not to wave flags or overpraise the heroism of the people of the United States military–I don't think anyone could underestimate the integrity of our people in the armed forces. I wish simply to describe the feelings of a dog-faced World War II infantry soldier operating under severe combat conditions.

Many people have asked me what it was like to have comrades falling wounded and disoriented around you in the heat of battle–to see some not breathing, lying bleeding with torn and mangled tissue. Sadly, your eyes soon become immune to the horror. To answer these questions would be like describing a strong odor, one with adjectives like pungent, rotten, undesirable. Only one thing describes the odor of war: it stinks.

I know accounts of the war have been written before by more talented authors with knowledgeable collaborators. Many military people have suffered more and experienced more combat than myself. My credentials are approximately six months of European combat, the Silver Star, Bronze Star, Purple Heart, Combat Infantry Badge and three European Battle Stars. I was drafted at the age of 19, before the legal age to vote. I was well trained in infantry combat.

I will never stop thinking of the thousands of men and women who gave their lives or parts of their bodies, including their central nervous systems, to secure victory in Europe. These people gave the supreme sacrifice so we can enjoy the freedoms of the United States. How lucky we are to be in this country. Even with all of its problems and controversies, it is far above any place in the world that I've been.

The account and actions are as I remember them–there is no fiction. The stories that I have written, including names or nicknames, are my best recollections.

I hope you can understand what goes through the mind of an infantry soldier from the accounts in this book. People in the front lines, especially those in the cold slit trenches or foxholes, often are the least informed of how the hostilities are going in the battle around them.

War is brutal and inhumane. It has a stinking odor you'll never forget.

The Heart of
the Infantry Soldier

Before this story of European combat, I wish to provide some insight into the infantry soldier. No matter where he fought, he knew the odor of war.

I have been told that in all American-involved wars in the past century, the KIA (killed in action) rate of infantry soldiers was 84 percent of all the military services. To attract young men into this branch of military services, we must offer extra benefits along with salary. I could name many such benefits, like no federal income tax. The infantry soldier of today is more advanced than the soldiers I fought with. Not only are more modern weapons available, but also military tactics that we could have only dreamt of back then. Certain traits, however, will remain constant. I have broken the five major elements of an infantry soldier into the following:

INTEGRITY

VERSATILITY

LONGEVITY

SENSE OF HUMOR

COMMITMENT

Integrity

The integrity of the infantry soldier is unmatched by any guardian of peace and freedom. He offers 100 percent of his physical and mental ability for long hours under arduous and brutal conditions. He truly is a lonesome doughboy, with his nearest human contact another lonesome soldier like himself.

The infantry soldier supports all facets of the military from the ships offshore to the armored columns of tanks and self-propelled cannons and Air Force armada. All aspects and weapons of military support come down to the lonesome doughboy... the single infantry soldier.

There never has been enough said about the single infantry unit. The role he has and continues to play in combat contributes much to the success of our forces. General Patton was famous with his armored divisions in Africa, Italy, France, Belgium, Luxembourg and Germany. He went nowhere without regular infantry, airborne infantry or armored infantry incorporated into armored divisions. Infantry units were very important guardians against single enemy infantry soldiers using anti-tank weapons.

VERSATILITY

Versatility was a common trait shown by American troops in the European Theater. It has been said that the American GI won the European war by his ability to improvise. Again, the single-unit infantry soldier demonstrated the most versatility in improvisation. It may have been because he lived afoot with only what little he had on his back and the gray matter between his ears to keep himself healthy, warm and safe. He was always on the alert to copy any good improvisation of the enemy (e.g., insulating boots and sleeping bags with soft straw or cardboard and digging into cow or horse manure for protection from the severe cold and wind).

The men in our squad designed a four-man dugout with heavy log coverings for protection. This dugout had an entrance and exit and was used when we were in a holding position in the wet, cold, snow-covered Ardennes. The German officers said that the American armed forces did not follow the book on military tactics. We did or used what was available at the moment. Right or wrong, we made it work. In most cases the improvisation worked out to our benefit, but no situation was cut and dried.

Improvisation was the key, especially for the enlisted men. Our commander in the Battle of the Bulge asked the question, "How can our troops survive in the sub-zero weather of the Ardennes? The conditions must be taking a toll on the enemy as well." The reasoning was true, but the enemy underestimated our troops. The answer to our commander's question is now in the history books. Our GIs were made up of all nationalities from all sections of America. Hitler called us the mongrel race. It turned out that this mixture of all races and experience made our guys more versatile.

LONGEVITY

Much credit for the durability of the infantry soldier can be given to the health inspections administered by our medics. Endurance training eliminated those who were missed by the medical exams. Some of those 30-mile forced marches at night were cussed and discussed in our training exercises. They seemed to be unreal and unnecessary—until the days of combat, that is. Somehow those same bodies would endure a double amount of stress in the same amount of time.

Taking care of the infantry body was an important step in our training, and personal hygiene was at the top of the list. I can remember that our teeth and feet always required the utmost care. The infantry soldier of today is going to be highly specialized with even more care given to his physical well-being. The longevity of the infantry soldier will be more important to the war department in the coming decade.

Sense of Humor

A sense of humor is very important to live the life of an infantry soldier. We had a big, strong southern boy from Alabama in our squad by the name of Leon Digby. He had a little more humor in him than the rest of us. I can still see him, when we were in a holding position, with a towel over his arm acting like a waiter in a large restaurant. He said, "Gentlemen, tonight's specials are canned hash covered with cheese, or cheese smothered in hash." Believe me, this made it a little easier to tolerate the freezing weather and incoming artillery.

The infantry soldier needed humor almost as much as his daily bread. One smile or the wink of an eye was worth a thousand words.

This is probably a combination of the other four elements that make up an infantry soldier. He meets the enemy one-on-one, eye-to-eye. The doughboy is trained to eliminate the enemy and hold the ground he occupies. In the immortal words of General Patton, "Kill those Kraut bastards and don't let the sons-of-bitches kill you."

I have seen infantry soldiers give the supreme sacrifice of war—their life or body parts. I've seen them out-survive the enemy in the brutal and freezing conditions of the Ardennes. This is commitment with a capital "C." This is war at its ugliest. I hope, in the future, our infantry soldiers will be used only in the defense of our country, and not for policing other countries.

ODOR *of* WAR

Chapter One
TRAINING IN CALIFORNIA

Many of us had been through basic training at least once, some twice, and we were at it again. We had been gathered from the Army Specialized Training Program (ASTP), which I was only in for six months. I brought with me one-and-a-half years of university training in biology courses with poor grades. Deep inside, I wanted to become an animal doctor. My brother, Joe, was a large-animal veterinarian and was doing great in his practice. My name is Andrew Giambroni and, at that time, I was 20 years old—one month shy of my twenty-first birthday.

D-Day in France was successful. Ground troops were finding fierce fighting in the hedgerows of Normandy. I believed they were training us for a Pacific landing, but who knew? The groups they were picking from the Army select ASTP group were desperately needed to replenish the depleted infantry forces abroad. This group was made up of men from 18 to 23 years of age, with IQs of 115 or greater. They were of top physical condition.

These young men were kept in reserve by the Secretary of Defense and utilized the depleted universities of 1943 for training. Most of us were studying engineering-related subjects with plenty of physical fitness training thrown in for good measure. I had to take ROTC (infantry) training at the state university before I was drafted. This training was mandatory at that time and became very valuable later in my military career. Map reading and position layouts skills were especially helpful. I can say this to young men (and women) today: secondary education has become a necessity in this day and age. I know when I was young I did not give my all to furthering my education. I gave in to side attractions and was not willing to sacrifice my time to deliberate studying. I learned my lesson.

Part of our training involved going through the infiltrated courses of make-believe combat. We did this in the dark of night, in thick mud with machine guns firing 18 inches over our heads. I never did find out if they were using live ammunition. They planted and set off simulated explosions near and around the course. The remainder of the night was spent cleaning rifles, equipment and fatigues.

It was summertime and the weather was beautiful on the Pacific Coast. Friday night was coming and weekend passes were easily obtainable. At 1600 hours one Friday, me and my two buddies, Miles and Tony, took off down the coast hitchhiking our way to the small village of Oceana. Of the three of us, Miles was the oldest at 21. He was a very handsome lad at 6′ 1″ and approximately 180 pounds. He was from Penn and an outstanding athlete. Tony, from New York, was more muscular, about the same height as me, and weighed around 175 pounds. Tony was 20 and, like me, had been drafted at the age of 19. He was dark-skinned with brown eyes and a build like an NFL linebacker. He looked Italian, which he was. Then there was me, Andrew Giambroni, called Andy, at 5′ 11″ and 160 pounds. With my light brown hair and blue eyes, most people didn't recognize my Italian heritage until they heard my last name.

The weekend was going great until Miles got into a beef with one of the civilian farm workers' girlfriends. They had consumed plenty of

Photo courtesy of Tom Moore Photography

June 1944—Weapons Platoon F Company, 385th Infantry Regiment, 97th Division. Andy Giambroni back row, far right.

tequila and lime juice and I guess by this time everyone had had a bit too much of the juice. The three of us battled seven or eight of them, though it felt more like 15. The brawl was on and soon the bartender called the local sheriff to break things up. He was near enough to respond quickly and got things under control. We all looked pretty rough. I had two swollen eyes that I could barely see out of; Tony and Miles were in similar shape. The sheriff took us to a USO where we could bed down for the night and hope to live another day.

We were late getting back to camp and were officially classified as AWOL (absent without leave). Our PFC (private first class) ratings were taken away, but they were looking for at least three infantry 745 riflemen from each company and, you guessed it, all three of us made the list. The Army needed trained military replacements. Our orders were cut and we were given three long train rides across the United States to the Camp Mead, Maryland, replacement depot.

We were called into the orderly room as soon as we got to camp. The captain said he wanted all three of us to go to Fort Benning, Georgia, to take paratrooper jump school training. He said it would take at least eight weeks for us to qualify as paratroopers. Miles and Tony immediately signed up, but I told the captain that I got dizzy standing on two bales of hay and I'd take my chances on the ground. My orders for transfer to the final departing zone were a few days off, but Miles and Tony were packed up and gone the next day. I was all alone and a long way from home for the first time in my Army career.

I wasn't upset about going overseas; in fact, I was looking forward to my first visit to Europe. We had two days of orientation, medical exams and vaccinations, then were given a weekend pass. I hooked up with two other overseas riflemen and we hitchhiked to Washington, D.C. We saw the White House and the Capitol and stayed overnight at a USO. We got back to the base on a Sunday to hear the rumor that we were shipping out the following Tuesday morning to Camp Shanks, New York.

The rumor was true. Monday morning we had final inspection and Tuesday we were off to Camp Shanks, the final departing port for overseas troops. I never thought for a moment that I would not return from the war. The guys over there were already doing a great job overcoming the enemy in the hedgerows of Normandy. I figured I could get there in time to celebrate victory with the Allies.

Chapter Two

SHIPPING OUT TO ENGLAND

Camp Shanks was great. The food was excellent and we had entertainment most every night and regular weekend passes to New York City. I thought if I could only entertain–play the piano, sing or something–I could have stayed there for the rest of the war. We were nearing the end of August and the newspapers were saying we were winning the war in Normandy and on the Brittany Peninsula.

I was issued long underwear to take with me. These underwear had a strong odor from the chemicals used to treat the fabric. They were to be used if the enemy used gas warfare.

The first weekend of September we were told to be ready to load up on a troop ship at a moment's notice. Eight days later, approximately 5,000 replacement soldiers, including me, were loaded up on a British luxury liner and headed to the war. I was told this was a very fast ship, and it was true. Even zigzagging without an escort across the Atlantic, we made it to England in 10 days. On the last two days of our journey, we were escorted by British cruisers.

During our trip, I spent most of my time down on D deck except for the two times a day we were allowed on deck for 45 minutes to an hour of fresh sea air. The British were damn good soldiers and sailors, but their cooking left a lot to be desired. Menus consisted of boiled carrots, cabbage, beef and, mind you, tea, not coffee. Everything was cooked in boiling water, and even though it was only a week or so, it was more than most GIs were used to. We bathed in saltwater and rinsed in rationed fresh water. We slept on lifesavers on the hard steel deck, which made the 10 days seem even longer.

Liverpool, good old England, looked good, but the sun was not shining. From the deck I watched them unload large nets of C & H sugar from the States. The tea they served in the Dungeon Hole had lacked even a teaspoonful of sugar. It was great to have ground under my feet and a bunk of my own to sleep in.

We had arrived at one of the largest repo depots in England. Here we were issued our orders to riflemen. We were handled in groups of our MOS (this was the code for jobs in the Army). I was in one of the larger groups—Infantry Riflemen MOS 745. I was officially getting overseas pay. I got a brand new M1 Gerand, one of the most outstanding rifles of World War II. My rifle was handed to me, full of cosmoline grease, still wrapped in brown butcher paper. We were given only soap and water to clean these rifles up enough to stand a stiff inspection by chow time. I always wondered why we were given the hardest way to accomplish cleaning a weapon that would be our life insurance policy for the remainder of our tour.

While in England I had a chance to take in a dance at Manchester. This was put on by the British USO. It cost one flourin but you got all the watercress sandwiches you could eat. We also were given a cut rate on the train ticket.

One morning after calisthenics I was called over the camp microphone to report to the company orderly room. The Captain there asked me if I would like to transfer to another part of England to train with the paratroopers. My size and age must have fit the bill for a paratrooper. Once again I turned down the offer—I was willing to take my chances across the Channel.

The English had already suffered greatly from the war. The harassing bombing had taken a sizable civilian death toll and many had lost family members. The Yanks were a welcome bunch. The English treated the American service people with great respect. There were the many WACs (Women's Army Corps) from the US who handled several non-combatant jobs with great efficiency. The US Air Force also was here in great numbers. They were overwhelming Hitler's great Air Force.

The German bombing of England had come to a halt and the Americans and Allied Air Forces had taken over the sky. They were taken off in "Armadas"—hundreds upon hundreds of B-17s and B-24s—on bombing missions to the Fuehrer's Deutschland. They destroyed many structures—railroads, trains, factories, ammunition plants, bridges, roads and seaports. How the German military could survive this punishment one could only guess.

The German Ack Ack had taken a considerable toll on our planes, aviators and Air Force personnel. The Germans were surprised that the

US Air Force would fly into the defense they had set up. I am still amazed that the German ground troops and armor could be such a tremendous force after the Allied Forces' backup destruction. We wondered how the enemy could keep fighting. We all thought it was just a matter of time before they threw in the towel, but they were probably afraid to stand trial for the atrocities we were hearing about.

Paris was liberated without a shot being fired, and the fighting continued in pockets throughout Western France. The Brittany Peninsula had been taken by my future outfit, the 6th Armored Division, a favorite of General Patton. The 50th Armored infantry was attached to this unit. They were the main aggressors taking the city of Brest at the point of Peninsula. Everyone in England was saying that we had the Krauts on the run. It wouldn't be long, we believed, until it was over. In this environment, I was ready to cross the Channel at any time.

Chapter Three

CROSSING THE CHANNEL

The moment finally came to cross the English Channel. It was late September and weather conditions were still okay. We boarded a good-sized ship at the port of Southampton. We had full field packs on our backs and carried our rifles at the sling. We were told to unbuckle our helmets from under our chins so we wouldn't jerk our heads off if we fell in the water.

We moored just off the shore of Omaha Beach. Daylight was seeping over the side of the ship into a personnel landing craft when we disembarked. We climbed down an iron fixed-rung ladder—very different from the rope ladders we had trained with. The ship's rocking motion gave us some thrills, especially packing 50-pound field packs and a slung M1 rifle. Surprisingly, with so many of us crawling down the side of that ship, I didn't hear of anyone falling. When the landing craft was full, we headed out to Omaha Beach. They pulled right up on the beach and dropped the bow. I don't think anyone got anything more than wet boots. This was the same beach where, more than three months prior, our boys had demonstrated their fighting ability to the great German military power and pushed it back.

We spent a few days at the beach helping to unload and ship supplies to the front. Gasoline and ammo had top priority. The heavy artillery ammo made us work hard and fast. I remember those few days were long and difficult. Our meals at the beach consisted of hard tack biscuits, Spam, beans and coffee. We dined in our little pup tents, two to a tent.

After a few days, more replacements came and took our place. We were loaded into crowded two-ton Army trucks that had no hoods. We traveled at night using only flashlights to guide the trucks down the long, narrow roads. We were lucky, the weather held out clear and dry until we reached our destination in the central part of France. This repo depot was on the outskirts of Toul, France. The rain came down in buckets creating a thick, sticky, deep mud.

My partner and I laid newspapers down and slept in our two-man pup tent. The second night at this depot, the majority of the new replacements developed the "GI Trots," known as diarrhea in civilian life. Climbing out of a warm sleeping bag, finding your boots in the dark, and dashing for the slit trench latrine made you glad to have long underwear on. Sometimes, however, the long underwear did not make it to the slit trench. Our boots were a big concern in that deep, sticky mud. Many times you would arrive at the trench with only one boot. I would have to say, to this day, I've never seen so much rain or stickier mud than in France.

We were on some of the same ground where World War I had been fought. In the deep trenches, those dog-faced infantry soldiers of 1917–18 had to be double tough. I read that the soldiers laid in these trenches for many days, badly wounded, with very little medical help. Many of these veterans had their limbs amputated because of gangrene. Without amputation, they would surely have died.

My partner and I spent a lot of time in the pup tent trying to stay dry. We read a lot, especially letters from home. I bet I read the letters in my shirt three or four times during that time. At night we could hear the sound of a single plane flying overhead. We were told it was an enemy observation plane. Between the darkness and the rain, I don't think he observed very much.

The next morning roll call was held in the tents and we were assigned truck numbers for departure. We scrambled to get our tents down, put our packs together, and grab our rifles, which were surprisingly dry. My tent partner was assigned to another truck so we shook hands and said good-bye. Our truck driver said we were headed for the 6th Armored Division, less than three hours away.

The weather finally cleared. We traveled by way of the big trucks until we reached the 50th Infantry Battalion, Company B of the 6th Armored Division. There were four of us and we were at Company B headquarters on the outskirts of Nancy, France, a disputed country of Lorraine. This was where Joan of Arc fought some of the battles of Christianity.

Company B headquarters was situated in a large, old chateau. The building had been vacated during the war and many of its outbuildings were being utilized by Company B troops. The chateau also was used as housing for the officers, first sergeant, company clerks and the headquarters squad, including the communications sergeant. Some of the 4th Platoon and the anti-tanks squad were here and so were we, at least until we were assigned. We laid out our bedrolls and set up in the attic of the old building.

Captain Frederick Silver was the new company commander, a replacement for the company commander who had been wounded in the Brittany Peninsula campaign. Captain Silver was 29 years old but the long lines on his face made him look a lot older. He had a pleasant smile and a tiny William Powell mustache.

The whole 6th Armored Division consisted of trucks, half-tracks, tanks, TDs (tank destroyers), mobile artillery and scores of weapon carriers and Jeeps. The 3rd Army, led by General Patton, was waiting for supplies. This flamboyant officer used rough language, but he was effective and given top priority for supplies, especially gasoline and ammunition. I saw General Patton in person only once. He was in a Jeep in the French city of Nancy. Patton was famous for his wartime quotes. At Nancy he said we were going to "take Metz if it takes a ton of dog tags." Easy for him to say—it was our lives on the line.

The German Army was going to battle at Metz in the Lorraine Country of France. The Rhine River could be their last defense. Back at camp we four replacements were given daily jobs. I was working with the ordnance crew. We repaired guns and received ammo and spare parts for the equipment. One day Captain Silver called me into headquarters and asked if I had any experience with military mapping. I had some experience from my training at the state university. I was immediately moved out of the attic to headquarters squad quarters. While I missed working in the ordnance where I got to fire many of the infantry weapons, especially the 50-caliber machine guns, I was glad to get out of that dusty attic. Each night we had to zip up our sleeping bags with only our noses sticking out to avoid the pack rats that ran rampant through the attic.

"Little Al," the guy next to me in the attic, was assigned that day to the 4th Platoon. I believe he was a rifleman and was used as an ammo bearer for a machine gun squad. The 4th Platoon was a heavy weapons platoon and included the M1 light machine guns and anti-tank cannons. I was going to miss Little Al, but we were both glad to get out of the attic.

Being in the headquarters squad of the company, I probably had a little more knowledge than the average GI as to what was going on in the conflict. Our primary source of general war information, outside of rumors, was the Army newspaper, *Stars & Stripes*, which came with mail call whenever we stopped. At Nancy we had hot chow three times a day and time to write and receive mail. This was where I first became familiar with *Stars & Stripes* and the cartoons of Bill Mauldin, who later became a well-known comic cartoonist. He put together the best portrayals of the dog-faced infantry soldier I've ever seen. It's hard not to laugh when you see yourself in the cartoons.

At this time we were told that the complete garrison of the Army, Navy and Marines was made up of approximately 8 million soldiers and that only a very small percent of these fighting units faced the enemy one-on-one. It was at Nancy that I learned what a big job it is to fight a war. It takes a tremendous amount of coordination to move, supply and resupply a fighting unit. The more units into battle, the more personnel is needed to coordinate the supplies. To fight a war and win, I found out, the guy who had unlimited reserve units with unlimited supply of fuel (gasoline) and ammo (artillery ammo especially) was odds-on favorite.

We were holding at Nancy waiting for the resupply of everything including replacements, ammo, fuel, gasoline and the bare essentials. More divisions were landing in France and heading to the front. Patton was like a racehorse at the starting gate, frothing at the mouth and wanting to get back into the race. During the two weeks we spent in and about Nancy, I became one of them, a member of Company B. I felt that I belonged to this outfit. The guys really impressed me with their abilities as infantry soldiers. They (the enlisted men) did not talk about their part in taking the city of Brest on the Brittany Peninsula, only to say that they had suffered several casualties, including the company commander.

These guys taught me everything about combat, especially the weapons we used. When I say everything, I mean everything that made them successful and kept them alive. I was packing a carbine now that I was in the headquarters squad. They showed me how to file down the sear on the M1 carbine to make it fully automatic of the 15 rounds in the clip. We kept this carbine in the track, but I know I never used it. I turned in my M1 Gerand, a great rifle especially on distance targets of five or six hundred yards. The communications sergeant taught me the basics of using a radio and the field telephone. The correct procedure had to be followed at all times; many lives were depending on communication between the units.

The armored infantry moved a great deal by half-tracks. It gave us more time to take care of equipment. It seemed you were constantly fighting for time to clean your rifle, brush your teeth and put grease on your boots. We used up a lot of the GI boot grease in the wet mud of France.

Chapter Five

METZ—THE LORRAINE COUNTRY

It was almost November and we were ready to move out to Metz, France, the German stronghold. I was told the city was built like a fortress. It even had underground entrances and exits to fit the situation. Most people in the Lorraine Country spoke German. This area of France had always been disputed between the French and Germans. The German-speaking civilians in this area of France had us confused—we hadn't read our history books.

It was already into the second week of November and we were scrambling to catch up with the rest of Patton's 3rd Army. It was cold and rainy. Headquarters squad was assigned seats in the first half-track. The armored infantry rode to keep up with the tanks and TDs. We were there to protect them from panzerfoust and the single enemy infantry soldiers firing them. Armored infantry also was supposed to protect against incendiary bombs thrown by enemy infantry.

I rode in the front compartment of the lead half-track with a turret-mounted 50-caliber machine gun and the driver. The Captain (company commander) also rode in front when he wasn't in the back with the communications sergeant on the larger radio. Often the company commander would be outside in the open, directing traffic. Captain Silver with his shoulder holster and his .45 automatic pistol was easy to pick out of the crowd. He could use that .45 with great accuracy, shooting many tin cans off of fence posts, while the rest of us were lucky to hit just the post. Captain Silver seemed like he was always thinking of what was around the bend. He was a medium-sized man who was always in motion, even when standing in one place giving complicated orders.

The rest of the crew in the back of the half-track were 30-caliber machine gun squad and ammo bearers, the communications sergeant and his assistant. The truck was full of many 110 and 130 wire spools, field telephones, radios, batteries and spare parts. The large radio and

Members of the 1st Platoon.

Members of the 1st Platoon. Note the German watches (Jerrys) and the German pistols.

Second Platoon Sergeant Mule.

*Andy Giambroni standing in the turret of a
50-caliber machine gun mounted on a half-track.*

communications sergeant in the rear seat made for a very crowded ride. The names of these guys are hard to remember, but their faces remain with me to this day. Some of their nicknames are pretty hard to forget. The communications sergeant with his bald head and soup-strainer mustache was called "Pappy." We had formed a family in this half-track. The first sergeant, company clerk, kitchen crew and support groups followed up the train of half-tracks behind us. As far as I knew, we were a reserve battalion, in a reserve division, as we took off to conquer the Lorraine Country.

It was the second day out when I heard and saw artillery shells exploding in the distance. This was enemy artillery. I had seen two or three dead German soldiers lying alongside the road with their destroyed weapons not far from their bodies. They were right there where anyone could just run them over, but our track driver avoided them even though they were our enemy. By the third day the artillery bursts were a lot closer and I saw my first dead GI. He was lying at the edge of a village with his shelter half over his body. His bayonet and rifle were stuck in the ground nearby with his helmet hanging over the butt of his rifle. This is war, but I thought, how rough can it get? I had hot chow every night and, although the nights had turned cold, I was warm in my GI sleeping bag.

The following evening Captain Silver instructed me to get some maps and colored wax pencils together—he and I were going to a battalion meeting. It was at this meeting that I first saw the brilliance of Captain Silver. He was not of a military academy, I understood; like many officers, he had come up through the ranks. He certainly did his country well. He knew the terrain and roads leading into the next few dorfs (villages). I was surprised to see majors and colonels asking him questions that I would have thought they should have known. Later I wondered if they were checking to see how much of the situation he had studied.

Early the next morning the three infantry battalions were in position. Our battalion (50th) was the lead battalion of operations that day. We were to point and hold ground while our tanks fired into any and all fortifications. This little dorf yielded little or no resistance. The rest of the German infantry were left to live another day. My job in this first day of combat was to give overhead fire from the 50-caliber machine gun on the track. I was not to fire until I heard small-arms fire from either direction and then I was to cut the hell out of any observation areas.

When I heard some of our boys firing light machine gun fire, I ran off approximately two belts of ammo at a church steeple and a long barracks where I had seen some scrambling. Cease-fire was called to the platoon leaders over the radio. The 1st and 2nd Platoons came out shortly, marching approximately 60 prisoners of war (POWs), but no officers except for some non-coms. "Blackie" of the headquarters squad, so nicknamed for his dark hair and complexion, was pushing a male civilian along with his bayonet. He said, "This guy only speaks German, I think he threw his uniform away." I told him what we had discovered, that the man was French and that the people who lived in the Lorraine part of France spoke German or French or both. I told Blackie to let the man go before he took the bayonet and shoved it up his ass.

After this first battle, I still naively thought that if this was war, it wasn't too bad. If we had any casualties that day they were minor. This was the first part of November and little did we know that the enemy was fighting a delay action into the same valley and river. The Germans were gathering their many crack divisions for the December break-through in the Belgian Ardennes.

The fighting became more intense the closer we came into the vicinity of Metz. This is where the German Army gave us a great display of artillery and motor fire. The Krauts were as good or better than any other military in the handling of artillery or motor fire. They had plenty of experience and training with these weapons. We kept pushing on around Metz, bypassing pockets where we encountered strong resistance. These areas later surrendered.

November turned colder and the rain seemed to come every other day. All of our vehicles—tanks, half-tracks, trucks and Jeeps—were having trouble getting through the deep mud. Supplies were getting harder to come by, especially gasoline and ammo. Weather conditions suppressed air support. It was our own artillery that kept the enemy off balance and seeking less vulnerable locations for defensive action. We moved at night with our columns strung out for many miles. This made us easy targets for enemy artillery and caused several casualties in our unit. Sometimes just standing still in the half-track with artillery shells exploding up and down and on both sides of the road would have you saying prayers to yourself. You prayed your track wasn't scheduled for the next direct hit. Your neighbor was probably praying the same thing. This happened many nights and would wind up in the early morning hours.

Our feet ached from the cold and the cramped conditions in the half-track. Food did not seem important even though we had K rations tucked in our jackets. We were averaging one-plus dorf each day. Division head-quarters were alternating our battalions and companies with ongoing armored tanks. Tanks were Sherman-type with 75mm mounted guns. Later on these were 90mm. The TDs were light and fast and a great weapon with the 90mm mounted. The support we received from the artillery units was invaluable. Without it, I couldn't imagine our chances of gaining ground in the sea of mud.

The enemy was stepping up its resistance. They were using the 20mm Ack Ack, a weapon that caused us many casualties and destroyed our radio and some other equipment. It was especially dangerous in the wooded areas where the projectile would explode and throw shrapnel in all directions.

The enemy were not surrendering very easily any more. The young SS troops would take a heavy mortar and artillery pounding and come blazing back with automatic weapons, including light machine guns. We got pretty close. We were throwing grenades at one another when two Krauts came out with their hands in the air and waving the white flag. One was shot and killed by someone in the 1st Platoon. We all hollered "Cease fire!" and in a few short moments approximately 30 Krauts surrendered. I said to the 1st Platoon sergeant to tell his men never, never to shoot someone who wanted to give up. I knew that many thought that Krauts were Krauts, but if they didn't surrender it meant more who would keep firing those ugly weapons they had. I have no knowledge of anyone in our units ever again killing anyone if he wanted to surrender. This was true even after we heard of the ugly massacre of the American infantry in the Ardennes.

Radio contact was our best means of communications. Hand-held radios were probably the best asset to communications, but in the cold, damp mud-caked area of Lorraine, they were called everything but a radio. Our communications sergeant rigged up one half-track radio on a backpack that was connected by pigtail cables to another backpack with batteries. This contraption weighed approximately 35 pounds but was connected easy and fast when Captain Silver needed to reach platoons, other companies, artillery units or whatever. Two infantry soldiers would pack this radio combination for many miles. It seemed we were always in search of fresh batteries which our communications sergeant always made good. Captain Silver's call sign, as you probably guessed, was Hi Ho Silver—send your message and over.

By early December we had reached some open country on the western side of the Saar River. The Captain said we were going into a holding position. The company commander's orders to the headquarters squad was to make sure all outposts and platoon sergeants had telephones in working order. To do so, our squad had to string out 130 wire to all platoons connected to the company CP (command post).

Enemy artillery fire was constant but every now and then we would have breaks of up to an hour at a time. Soon enough, though, the artillery would come in again with shells of the 105 howitzer variety. Our squad would time the report of the enemy gun until the projectile landed and exploded. We would then take a back angle of the shrapnel area, using the river as the base. This information was relayed by Captain Silver to our artillery fire center for response. They commended Silver and his men for quieting some of the German artillery batteries by this practice. The Captain told the men that it was a wonder they didn't gather enemy artillery shells in their pockets when they were getting the angle of the spent shrapnel lines. The holding position seemed more like the type of war I would rather fight.

Christmas was around the corner. We knew it was December when the packages from home started arriving. It was still wet and cold but we were eating hot chow and the diarrhea plague was clearing up. The packages for the men who had been killed or wounded were opened and the clothes, such as wool socks, scarves and long underwear were divided among the squads. The cookies, smoked salmon and many kinds of candy were a real treat. All valuables and letters were returned to the senders. We knew it was a federal offense to tamper with the US mail and packages, but we felt Uncle Sam and especially our comrades would understand.

Around the 21st of December a complete infantry division arrived with equipment and men wearing snowpack boots and long trench jackets with hoods. They really looked sharp. We were being relieved— the whole 6th Armored Division was going back to Metz.

Chapter Six
CHRISTMAS IN BELGIUM

Metz had been cleared for a month by the time we got there. It was the 23rd of December. That evening the cooks prepared a Christmas dinner of canned turkey. They told us General Patton's 3rd Army was needed in Belgium. I was running a temperature of 102 and the captain sent me to the aid station where they gave me sulfa pills and kept me on a stretcher covered in blankets to sweat out the chills of the flu. I slept and sweated it out for 24 hours, and when the medical captain took my temperature again it was normal. He asked how I felt and my response was, "Weak as a cat." It was then I was told that Captain Silver wanted me and three men from the company to go by ambulance to Arlon, Belgium—a distance of 40–50 miles.

We arrived in Arlon on Christmas night and the Belgium civilians there were sharing some Christmas cheer with the soldiers of Captain Silver's Company B. The mayor of the town gave an emotional speech in English. These people also had been run over by the Nazi regime. They told us they would pray for us and for our families, that we would bring freedom to the world once again.

That very night Captain Silver, the executive officers and I went to a battalion meeting. The 4th Armored had broken through to the surrounded 101st Airborne Division. The paratroop division had held their own in a standoff with the mighty crack divisions of the German Army against overwhelming odds. They were now isolated in the Bastogne area of Eastern Belgium, the entrance to the Ardennes forest mountains. This was the worst winter that the Ardennes had experienced in 50 years. I might say, 50 years later, they have never had one so bad since.

The 101st Airborne Division is one of the outfits that demonstrated great fortitude during this conflict. They held position many times against tremendous opposition. The 101st did not jump into the

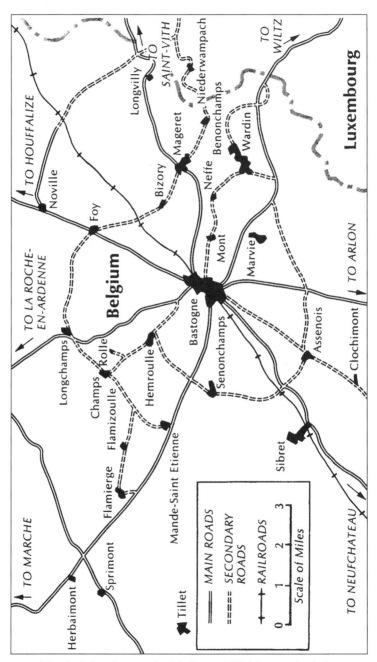

Map depicting the area the 6th Armored Division was headed to.

Bastogne fight, they came in trucks and vehicles of all categories. They fought the Krauts hand-to-hand and would not surrender.

The German command had encircled the 101st and sent in an enemy contingent under a white flag to encourage General McAuliffe of the 101st to surrender. According to accounts, the Germans told McAuliffe that if he had a negative answer they would annihilate the Americans down to their very last man. McAuliffe's one-word answer was written on a single sheet of paper: "Nuts." This was one of our youthful sayings at the time: "Nuts to you if you don't like the situation." It must have taken the enemy all afternoon to figure out General McAuliffe's reply.

To this day many Belgium people have great respect for the "Screaming Eagle" 101st Airborne Division. There is an area in Bastogne where the citizens have planted groves of trees with brass plaques carrying the engraved names of infantry soldiers.

The weather was growing colder every day. Keeping our hands and feet from freezing was a full-time job of its own. The Germans were like a sleeping lion that had been kicked awake. This was the last round of the fight and they were going for the knockout. I was becoming increasingly scared with our situation. War didn't seem as easy now. Dead GIs lay in the snow, frozen solid and waiting to be picked up. The intense incoming artillery also had prevented the Germans from picking up their dead, and the frozen curbs were lined with the corpses of the enemy.

When we moved into a position of attack, our infantry and armor were spread out. The small village was Bizory (it's part of Belgium now). Almost all of the civilians had left and fled to the west. The roads to the village were lined with stone homes, some with long stone barns attached or in close proximity.

Many years later I found out that a 17-year-old Belgian boy and his father had slept on top of potatoes and turnips in the cellar of one of those barns during the conflict. The potatoes and turnips were fed to the cattle that had been bred by their family for generations. These are the Belgian Blue breed of cattle that are now being shipped all over the world. Several head of these cattle were held in the long stone barns of the family farm. The only cattle this family lost were the ones the Germans took for food.

*January 1945—Aerial view showing armored and infantry of the 6th Armored
Division advancing through the snow in the direction of northeast Bastogne, Belgium.*

*January 1945—View looking north from Bastogne, Belgium,
after the Nazis were driven from the besieged city.
Heavy vehicular traffic rolled through this area constantly.*

This 17-year-old Belgian stayed with his father for more than a month without heat, using only blankets to keep them warm. They saved this valuable breed of cattle with their guts and commitment. I met this young man many years later when I visited Europe. He was 62 and I was 66. I don't think either one of us could stand those conditions today.

They speak French in the eastern part of Belgium. My wife and I had an interpreter with us when we visited the Bastogne area. This young man, now in his seventies, is Jules Schumer and he had grown up in the Bizory-Bastogne area of Belgium. When my wife was telling him that her husband had fought the Germans in this spot on the outskirts of Bizory, Jules wasn't too impressed. His family had met many Americans who had fought in this area, and the Bastogne monument and museum were only one mile from his farm. I was standing a short distance from them when I heard their conversation. I asked him to come to the edge of the road in front of his farm where I pointed to a large shell hole that several GIs and I had jumped into when we heard incoming artillery shells. The large shell hole was at the entrance to his farm, not far from the barns where the cattle were kept. With this, the expression on his face changed and his eyes began to sparkle. He told the interpreter to have my wife and I come over to his house for coffee and cake.

We visited with Jules and his family for some time. He called the German soldiers the Bosgh. The Americans troops sure did eliminate a large number of the Bosgh, he said. He expressed the gratitude of the Belgium people to the American soldiers. If you are ever in Bastogne and are interested in the beautiful Belgian Blue cattle, Jules Schumer is a 10-minute ride out of the city of Bastogne.

We now return to our story of the sub-zero winter of 1944–45. The weather was enough to contend with, without fighting a brutal war. We dug foxholes and an observation dugout in short order. We had run telephone lines out to each of our platoons and one to the FO (forward observation) officer who was living with the tankers. K rations were the best we could get to eat. Our little Coleman stoves did heat some coffee, tea or soup. We dreamt of ways to keep warm. The German Army was working the ground hard and we had to stay alert and be at the ready. If they didn't come at us, we were going after them. We kept weapons, ammo, extra pairs of gloves and wool caps under our helmet liners and, sometimes, a K ration inside our jacket.

The main drive of the enemy followed an arrow point from the city of Mageret. The way I understood it at our battalion meeting, the 6th Armored Division was like a tailback, poised to get through the hole and go for it. Our area was about three or four kilometers directly east of Bastogne. It took us three or four days to get into position. We were in the small farming community of Bizory. It was only two or three miles the way a crow flies out of Bastogne. The date was the 30th of December. The enemy artillery was unlike anything I'd heard or seen before. Years later I learned that the Germans had laid a 20-minute barrage of 500 artillery shells in our area. There had been some serious fighting with numerous dead German soldiers lining the small country road to Mageret.

Chapter Seven
MAGERET: FIRST ENGAGEMENT

The snow was deep with two- and three-foot drifts in a lot of places and the temperature was in the sub-zero level. The clouds and overcast had a low ceiling which meant our air support could not help us. Early in the morning on the 31st of December we took off on the two kilometers' journey to the town of Mageret. We had encountered strong resistance before, but nothing like this. The enemy was well fortified and had machine guns and 88mm direct-fires in place. Our artillery had blown out most of the buildings in this little town.

The enemy had dug in near the bottom of a sloping hill. They were camouflaged in white against the snow, but we knew where the automatic weapons were, especially the machine guns. Our slow-moving tanks were getting hit hard and moved behind anything to keep from getting hit. I stayed near the tanks for their protection but far enough away to keep from the direct fire of the 88s.

We lost the first man of our squad that day. I only knew him by the name of Derner, but I can still see him with his big smile, smoking one of his crooked-stem pipes. Two of our squad were wounded with machine gun bullets in the leg. The 3rd Platoon leader, Lt. Green, had been wounded in the chest and shoulders. Captain Silver told me to get him back to one of the half-tracks. On the way back with Lt. Green, we were joined by numerous wounded soldiers being evacuated. With the enemy machine guns shooting and our own tanks spraying, the chances of picking up a bullet were pretty good. I was in the headquarters half-track dressing Lt. Green's wounds where I could see our task force falling back. We were to prepare a defense line ourselves. Company B had suffered a large number of casualties and we were short of man-power. A counterattack would be coming soon.

Every man was instructed not to take off his boots and to always be at the ready. The FO officer of the 6th Armored fire center took me and a list of maps showing what would be the likely avenue of a counterattack. He was a first lieutenant and sure knew how to lay out a pattern that would slow down a counterattack. The most likely area was our left flank. There was a long line of hedgerow and a line of trees that were great for the advancement of infantry.

The 2nd Platoon was dug into the fringes of a 200-foot stone barn protecting the area. This position also kept most of the 2nd Platoon out of the cold winter night. They had several light machine guns and mortars from the 4th Platoon. This setup had great visibility, but the two main listening posts were far out, about 50 to 100 yards. These guys had overcoats, wool blankets and a telephone. Seeing them out there with icicles hanging from their nostrils and ice forming on their beards made you feel lucky to get cover even in a blown-out building. They were out there for two or three hours at a time. Mule was the platoon sergeant of the 2nd Platoon. He was a good-sized man with a heavy black beard. Every time I saw him it seemed he had snow sticking to the edges of his beard.

The next day I discovered an old iron-rod bed with just the mattress on it at Company B headquarters in a small, two-story rock building. The civilians had left there when the fighting started. I looked at that bed and pulled off my boots, which had been on for at least three days straight, and fell sound asleep on the mattress. I was so exhausted that even the enemy shelling did not disturb me.

Our air support had visibility and had started bombing the enemy positions. The orange tails of the P-47s were a good morale builder for the troops. My lovely nap was interrupted by Captain Silver hollering my name. He came quickly up the stairs and his first comment was, "Damn, your boots are off." He told me to get them on and come downstairs. Sergeant Taffy and I were to take a telephone and a reel of 130 wire to the 2nd Platoon. Mule had told the Captain that the enemy was gearing up to come our way. The FO lieutenant was waiting for us to bring extra maps of the area.

It was nighttime but we had some moonlight to travel by. The 2nd Platoon was only a quarter-mile away but the distance seemed longer tromping through the cold snow. It was not snowing, and Taffy and I had all the clothes on we could carry. We slipped out as our CP lay out the 130 wire, the wire reel creaking as it went. Across the road was a large shell hole and Taffy and I had to jump in it when incoming artillery hit all around us. We made it to the 2nd Platoon with that squeaky wheel of wire and our hearts in our throats. Once there, Taffy, another soldier and I had to repair a blowout on a wire to the listening posts. It took three of us to do the job—one to get the two ends of the blow, a second guy to tie the wire in a square knot, and a third guy to wrap the splice with electrical tape, each one of us working with our gloves off.

Mule returned to the platoon and said that the enemy had stepped up their shelling. We were instructed to be awake and at the ready. The listening posts returned. Shortly we could hear movement coming our way. The FO, Taffy and I were on the telephone to our CP. He knew the coordinates of the avenue of approach. Within moments our artillery could be heard going over us. This was a good sign. If we had to move targets, we knew where our guns were. New Year's Eve was never lit up like that long field on the outskirts of Bastogne. Between the artillery, machine guns and mortar fire, the adrenaline was hopping in every dogface dug into that barn that night. The whole thing didn't last more than an hour—to me it seemed like forever. We took turns trying to get some sleep while artillery burst in the distance.

Early the next morning, Mule took a patrol to investigate what we had accomplished. He reported finding a musette bag that had been torn to shreds and said there were spots of blood in the snow. He thought they had taken their dead and wounded back toward the town of Arloncourt. There was spotty sunlight that morning. Taffy and I took turns taking off our boots and rubbing each other's feet. It was really cold and I knew the enemy was just as tired, cold and hungry as we were. Who thought up this B.S. called war anyway?

I can't remember it if was January third or fourth but that day I saw six of our Sherman tanks hit by 88s about as fast as you could count. It was right at daybreak and they hadn't had the chance to move back behind the ridge where they usually stayed during daylight hours. I'll never forget the sight of the turrets of those tanks flying 20 feet in the air. Many of the tank crews, being in close proximity to their tanks, were killed or wounded.

The fog and overcast weather was rising fast and visibility was in favor of air support. I understood we were getting a great deal of weather casualties—soldiers suffering from frozen feet, hands and fingers. I am sorry to say that our equipment—clothing, boots, socks, sleeping bags, wool underwear, gloves, stocking caps, and ear and face protection—was in short supply. Ammo, fuel (gasoline) and water came first. It was hard to believe that water was in short supply when we were surrounded by so much snow. The five-gallon drums of water were being delivered frozen solid. We soon found out how much snow to melt for one cup of coffee. The old black galoshes we had in France were as useless as teats on a boar. When replacements came into the company with the newer garments and snowpack boots they had to sleep with one eye open. To tell you that I never thought of taking my carbine and shooting my big toe would be a lie. In fact, I thought about it too many times. I can say truthfully, though, I did not know of anyone in our outfit who committed such an act.

While I don't consider myself to be very religious, I did pray a lot at that time. I had lost my father from a heart attack shortly before I was drafted into the Army. My dad was and still is my example. He was a very kind man who lived by the Golden Rule. Love and be kind to your neighbor and you will be rewarded on this earth was his credo. I really believe my dad could have made friends with a rattlesnake. I felt that I didn't know God well enough to ask him for any favors, but my dad I could talk to. I would tell my dad that if God was going to take me, please don't let him take part of me and send part of me home, take all of me and I will be with you to watch over the rest of the gang—two brothers, two sisters and my mother. I felt my dad was with me every step of the way, watching over me and guiding me.

A day or so went by—time didn't have much meaning then. We kept busy repairing phone lines. Our headquarters squad had lost one man and had two wounded. In some of our platoons the whole squad had either been killed or wounded. This was truly a bloody battle and there was more to come. The counterattack was halted primarily with our artillery. The loss of those six tanks and the evacuation of several of the men who were in or about those tanks was hard to take.

More replacements came in—how young they looked! Some of them had eaten Christmas dinner in England, others were volunteers from other units in England. Many of the new recruits into Company B were really young (18 and 19) or old (in their late twenties). The older guys were volunteers from other branches of the service. All of the non-com of

Company B gave them a short course on handling, loading and firing the many weapons. We took special care with the M1 Gerand because this was the weapon most of them would carry. The officers and platoon leaders, mostly second lieutenants, were very expendable. I never got to know them very well because they usually ended up dead, wounded or suffering from frozen feet. This was often the fate of the FO officer as well. When I went to the hospital in March of 1945, our Company B had two battlefield commissioned officers. The captain was looking to send more sergeants with combat experience to Paris for battlefield commissions.

We were holding our own against tremendous enemy power. The Germans were running out of fuel, especially for their large tiger tanks, and were having to ration fuel for the German Air Force. We had to keep a watch out for the enemy who were infiltrating our lines in American uniforms. Those Krauts could speak English very well, without any accent. We halted everyone at our guard posts asking for name, rank and serial number. We also were supposed to ask a question that any GI would know, like who was Joe DiMaggio and what team did he play for or who is Clark Gable and what is his occupation. Our company did not run into any of these spies that I know of but there were some caught by other outfits. We had to be doubly careful now. The enemy was desperate and would try anything—right or wrong. They were taking desperate measures. It was then that I thought regretfully about the gas mask I had left in France.

Our own medical garrison was well overworked, taking care of GIs and the wounded Germans. They must have taken care of as many enemy wounded prisoners as they did our own. Both men and women worked around the clock attending to the wounded military personnel. Years later I found out that many medical procedures were first utilized in this war. From the enemy our medical personnel picked up the intermedulary process of pinning broken and mangled bones. This was first noticed in the X ray of a German prisoner with an injured leg who had long pins in the marrow of his bone. It wasn't long before many limbs and parts of the body were saved and healed in position using this practice. This was a great medical breakthrough; it was just unfortunate it took a war to exchange such an idea.

It seemed that everyone had found out that I wanted to become a doctor. I don't think they knew that I wanted to be an animal doctor.

Even the medics would bring guys with a medical problem for me to look at. I always did something if I could. Our own medic was a truck driver in civilian life. He was always there when you needed him. He packed two large musette bags where he kept everything from treatments for the common cold to the itching of inflamed hemorrhoids. At this time, antibiotics were not the common resources that they are today. Even penicillin was just being tested by the FDA. Our own medic spent most of his time treating health problems and superficial wounds. Much of my time was spent helping him.

I did have enough spare time then to get off a letter to my mom. She always insisted that I change underwear after a shower. Boy, what she would think of her bundle of joy in two-month-old underwear. Bathing out of a helmet was the PTA method—privates, tits and armpits—if you could get out of the cold long enough to do it, that is.

January 1945—Belgian civilians leaving their homes in the forward fighting areas during the sub-zero weather to seek a place of safe haven in the rear of the lines.

Chapter Eight

One of the men from our squad who had been wounded was coming back from the hospital with new replacements. He was dropped off with his duffel bag at the CP, but before he could get inside an artillery shell exploded nearby and a small piece of shrapnel entered his abdomen. Within a short time he was in an ambulance headed for the aid station. The war was over for him; I never saw him again.

That evening I again went with Captain Silver to a battalion meeting. The orders were that Company B would head out at dusk down the long slope to the town of Mageret using the darkness as cover. The tanks and TDs were to follow us with heavy gunfire. The half-tracks were not being used but would be at the ready in case we needed them.

All of our equipment had to be checked, especially the guns. Outside at Company B's lead half-track, the men were getting the radios and guns in shape. They were having a problem with the bolt on the 50-caliber machine gun—it was frozen tight. We tried using light oil but it still wouldn't give. Leon Digby had a brilliant idea. Before we could do anything, he urinated on the bolt and it slipped free. From then on, we used a cup of warm water. That Leon, he was always thinking!

We knew we were in for some serious business when the chaplain arrived. He happened to be Catholic, but it didn't matter to us as we knelt in the cold snow by the altar on the hood of his Jeep. He said a prayer over all of us and offered communion to any of the Catholics present. He paused between placing the communion host on our tongues to warm his hands. I remember one guy telling me he really believed the Catholics were ready to die when they received communion. That's a strong faith, he said. Twenty-four hours later, many of those men were in heaven.

When we jumped off, Captain Silver wanted our radios on the back of the FO tank. He and I and two other men rode on the back of the tank. We started our advance after the P-47s and artillery gave an hour or two of fire to soften up the dug-in Krauts at Mageret. It was still light enough to see but darkness was coming fast. Small arms and machine gun fire were intense. Captain Silver was bleeding under the brim of his helmet from a small-arms bullet that had creased his forehead. I grabbed his belt aid pack and wrapped it around his head to stop the bleeding. I remember that I couldn't get low enough, hanging from that tank turret.

The enemy 88s opened up next and hit the tank to our right. They pulled over immediately behind some broken buildings with our tank pulling in behind them. The hit tank was on fire and the men were being pulled out as quickly as possible. The tanker took two of the men who had been burned real bad into the broken-down building. I grabbed the first aid kit from the FO tank and gave each man a syrette squeeze of morphine in the arm—I believe it was a half grain or so. They were suffering from shock and their faces had two- or three-inch-long blisters hanging down. The best I could do was to rub sulfa cream on their faces. Today I know that was the worst thing to use.

It was dark as hell but we did have the light of the moon to go by. Out in the darkness across the stretches of moonlit ground you could hear boys hollering for the medic. The Kraut machine guns were kicking up the snow in stretches. I never in my life felt so helpless. None of us stood a chance going out there to get anyone. There was a partial crew of tankers, two of our squad and the Captain. In what was left of the building two German soldiers had given up to our squad. I talked to one of these prisoners and in my broken German and his broken English asked if they would drag any of our wounded back to the building. I was hoping that the Kraut machine guns would not cut down any of their own. They agreed to try. They got out about 25 yards and came crawling back to the building. I was willing to try anything, but the Kraut machine guns spoiled any crazy plans I could come up with.

The Captain was talking to the FO tank commander. I knew they wanted to get out and flatten that machine gun. All of a sudden two guys of the headquarters squad hollered that Krauts were coming up the road. They were under 100 yards away and clearly visible in the moonlight. We lay at the corners of the flattened building and fired directly at them, knocking them all down. I knew then how it felt to be scared shitless. Captain Silver who was wearing the Joe McGee head bandage had his

.45 drawn. Everyone now had a sharp eye out. Silver told the tank crew from the FO tank to go straight, turn left and take the first road up the hill. He said if we moved fast enough we just might get out of there.

We needed help and our radio was not working. Silver had me and another guy, Red Bryant, climb up on the tank to see if we could get the radio to work. With Captain Silver hanging on the tank's turret and me and Red on each corner, the tank went downhill for a short ways, turned left and went wide open for 300–400 yards at the bottom of the town. I still was working with the handset trying to get someone to come in when the tank took a sharp left and started up the hill. There was a loud noise and a large flash and before I knew it I was on the ground rolling in the snow with the radio handset in my hand, my carbine slung over my shoulder. I could hear the Captain hollering my name as he was holding onto the turret. The tank was not crippled and I could hear it running to the top of the hill, approximately 400 yards away. We had been hit by a German bazooka panzerfoust, a weapon operated by only one man. I was lying off the shoulder of the road and had my carbine ready. I heard someone close hollering my name but I stayed there and kept quiet, thinking to myself, "If this Kraut is still close by then he heard the Captain holler my name. I'm not going to stick my head up to get it shot off." My heart was in my throat and it was going 100 miles a minute.

I was checking my carbine when I saw Red standing in the middle of the road. He had been blown off the left corner of the tank, the side the Kraut had shot from. He looked funny with his hair on his face and eyebrows burnt away. He kept calling my name until I told him to get over to where I was and get down. I knew the Kraut who had fired at us wasn't far away. Our helmets were still on and Red had his carbine. I heard the tank stopping at the top of the hill. As Red and I went up the hill on the road we fired a couple of rounds into every manure pile that looked like a dugout.

The Germans had been smart enough to use manure piles for protection and warmth. When we finally got close to the top of the hill the Captain hollered at us to stop shooting our guns and to look to the skyline. Approximately two miles away we could see lines of German foot infantry moving out. I will always remember what Silver said: "Those are enemy infantry and they're not doing close-order drill."

The gunner in the center of the tank had been killed. The crew had pulled him out and he was lying on the cover of the tank. We had a one-and-a-half- to two-mile trip to our CP in Bizory. Before we got started, Red went up about 150–200 yards to scout and came running quickly back. He said there was a Kraut in a hole not far from us. Captain told me to flank from one side and Red from another while he covered us with his .45. Red got to the hole first and jerked a rifle out onto the snow. The Kraut had his hands up and kept saying he was wounded. When we got him out of the hole we could see he had one leg shot up pretty badly. We had to help him up on the tank and took him to Bizory and the medics.

Captain, Red and I walked guard with the disabled tank. The panzerfoust had hit a little high, and hadn't crippled the tank's boogey wheels but had made about a three-inch hole on the side of the tank and caused a lot of shrapnel inside. At the CP in Bizory we were told that the FO tank was the only one to come out of that fight. Fifteen of our tanks, a whole company, had been knocked out. Another company of tanks and TDs were on their way to us. A platoon of combat engineers was sent in to bolster our beat-up Company B.

Early in the morning Captain got together what was left of our company, added the combat engineers and tank company, and moved back down into Mageret. The enemy had moved northeast to the village of Arloncourt, about three to four miles from Mageret. Arloncourt was in a valley surrounded by woods and had big open sloping hills that did not leave much protection. During the days, with the help of the combat engineers, we recovered some of our wounded. We also found more wounded Jerries, but the number of dead on both sides was greater than I had imagined. I really believe the Krauts thought we had more troops than we did and had decided to move to Arloncourt for a better defensive area.

Most of the day was spent rolling over dead Krauts and marking our own dead so they could be seen and picked up. German machine guns and direct fire had taken a big toll on our infantry riflemen. There were four dead GIs lying in the snow up at the area where we jumped off the previous evening. The enemy machine gun must have hit them when they first came over the brim of the hill. They were lying in an even line, 15–20 feet apart, face down. A silver shining watchband glittering in the afternoon sunlight off the left wrist of one of the men caught our attention. One of my guys said, "I better take that, he won't need it anymore." I growled at him, "No! No! That's bad luck. Leave him alone." I did not touch the man but studied his face. It was little Al who had slept in the attic in Nancy with me just a few short months before. I had thought he was in the 4th Platoon. His face was almost as white as the snow. He must have been hit in the head by a machine gun bullet. A large triangle of mucus and blood was frozen out of one of his nostrils. This was the face of the little guy who could smile and make your day. But no more.

Captain Silver in the new CP was now at the far east end of Mageret in a building that was still partly together. He had secured the town and even had hot chow brought down. We needed more replacements, but this was true of most of the companies in our battalion. Our company was still taking in prisoners, mostly wounded, some aid men and those who were carrying the wounded who could not walk.

I can still picture that night. One German soldier had had his face completely shot off. He was walking slowly with another prisoner leading him. The sight will stick in my mind a lifetime. He had no forehead below the hairline, no nose, no eyes–just bubbles of mucus and blood where the nose and mouth would have been. I can only say it was lucky that the severe cold had congealed the brutal remains of his face. I always wondered if his medic gave him morphine. He could have died from the shock alone. Our losses of both killed and wounded were great, but the enemy losses of just dead alone had been much greater.

The Captain had a professional bandage on his head and he had replaced his helmet with an overseas cap. He would eventually be chewed out for wearing just an overseas cap. Back at Nancy, France, I had known everyone by name or face. Now, a couple months later, we had 50 percent new faces. We needed more replacements, fuel and ammo. They all came bundled in big two-and-a-half-ton Army trucks.

Prisoners were still coming in but most of them were not wounded. They were stragglers who had been left behind and did not want to fight anymore. These were not the SS troops or the younger generation that had been brainwashed into fighting for the Fuehrer and Deutschland. Those young bastards were deadly with all the infantry weapons. In any village or town those young SS would hide in the ruins of the buildings so they could snipe and kill as many GIs as possible. Many times we caught them, but those daredevils always had an escape plan. Several got away. When we captured one of those SS troops he would just stare at you and smile a shitty grin. We knew better than to turn our backs on one of those bastards. He would stab you with anything he could get his hands on. Those prisoners were handled separately and would be taken to the rear under special guard. The GIs would take most anything off a prisoner or even a corpse. This was the price of war. Officer prisoners were a real prize because they carried a pistol and had many medals showing on their uniform.

Some of our lightly wounded went to the aid station and reported back to duty the next day. One soldier in the 1st Platoon had a bullet go through a K ration in his jacket, traveling clean through a can of cheese and just breaking the skin on his chest. This must have been a spent bullet from a long way off, but he was lucky nonetheless. While we were still looking for any of our wounded who were still alive, we went into a building where we found a blown-out window and two dead German infantry with their heads blown off. They must have been firing out of the window when a Sherman tanker hit the window with a 75 cannon.

I still couldn't get over the fact that the Germans had pulled out of Mageret when they really had us bent hard. If you saw the scattered dead German bodies around the battered buildings of Mageret, though, I guess it made sense.

I had never seen war at its ugliest until I was fortunate enough to survive those 10 days. It was hard to believe the monumental stand that the 101st Airborne had made until the arrival of General Patton's 3rd Army. They were trucked into the Bastogne area in a defensive holding position. They were surrounded by an overwhelming number of German infantry and armor. Enduring the freezing weather, with little or no anti-tank weapons, they depended on the GI bazooka, a weapon operated by two infantry soldiers. With these few weapons and the German Army low on fuel, the men of the 101st Screaming Eagles will always be remembered as one of the great military defenses of the European war. The effort and bravery of these airborne infantry soldiers is hard to visualize—they had to be double tough to stand the odds and the conditions in which they fought. They created an example for us all.

In the Ardennes, the armor was better suited for defense positions than for aggressive ones. We were now the aggressor. The German Army was taking advantage of the long sloping terrain of the forest. We learned very quickly to look for hidden armor protecting the open areas of the Ardennes. The enemy tanks were armed with 88mm direct-fire cannons. We knew better than to challenge their accuracy. We avoided these areas for the most part, but did get into these open shooting galleries from time to time.

I can still see Captain Silver studying maps of the areas. He had to put on reading glasses to study the maps. His beard grew really fast, but for the most part he kept clean-shaven. Silver seemed to always have a cigarette dangling from his lips and he would put a shot of cognac in his coffee to keep him warm, or at least that was his excuse. He was a loner and a quiet man for a person in command. He did not want to know people in his command personally and it hurt him deeply when one of our company was killed or wounded. The Captain never said this outright but you could tell when the 1st Sergeant gave him active roll call and he asked questions about every man. I saw Silver's eyes fill with tears after this last encounter at Mageret. After a few glasses of cognac and a pack of cigarettes, he was twitching his lips and rolling his shoulders. He then strapped on his .45 and I knew he was ready to get on with the business ahead.

Chapter Nine
ARLONCOURT

This encounter is singled out from many because it caused a great deal of casualties. We knew we had the great German Army retreating; they were back to playing delaying action. The Siegfried Line pillboxes and tank traps were not far away at the Our River, about 30 or so miles away through the Ardennes mountains. The Germans knew we couldn't move armor or infantry very quickly through these snow-covered forests. We were getting more benefits from our air power, though, and the enemy artillery slowed way down when the P-47s were around.

Most of us were still suffering from diarrhea. It was a challenge dropping your drawers to release your bowels in the snow. A guy would dream about the nice warm potty at home every time the urge set in. The little things in life become more important when they are not available. I tried not to think of the atrocities of war and how the human brain made allowances for such acts to happen.

At our half-track the squad had brought in a wounded German soldier. He was laid on the hood of the track. The engine had been running and the hood was warm. He kept saying *danke* (thanks). One of his legs had an ugly wound in the thigh area and he had on paratrooper boots. One of the guys in our squad said that the German soldier must have taken them off a 101st paratrooper. The GI said that he wanted the boots because he thought they would fit. I said it was okay, but that he had to give the German his own boots. He said no. I told him that I would reason with him only once. If he took the German's boots and left him with nothing, his feet would freeze and would have to be amputated. Then he (the German soldier) would be a major problem for everyone who attended him. Maybe, if he was okay, he could unload supplies in France and perhaps send me a pair of snowpack boots. My reasoning worked and he agreed. He understood that I wasn't a Kraut lover, just interested in being fair.

Everyone was convinced the war was about over now that the Germans were on the run. The enemy hadn't given me the impression they were giving up. They were still as deadly as ever and had come close to knocking us out.

That afternoon six or seven truckloads of replacements and supplies arrived. Our squad got two replacements, one from the Air Force in England and the other a young kid who had eaten Christmas dinner in the States. The guy from the Air Force was getting lessons from our squad on loading and cleaning the M1 rifle and other weapons on the track. He was an older man, in his late twenties. Someone in our squad said he had volunteered for combat. We needed replacements, but to volunteer for this kind of duty meant your patriotism was awfully high.

We were well into January of the new year. Arloncourt was a problem and even the dive bombers were having a hard time softening up the enemy position. I remember in France we had hundreds upon hundreds of B-17s and B-24s flying over to targets in Germany. Even with the bombing they took then, the Germans still came fighting back. The armada of planes sent out in October and November was so large that the lead planes were returning to England while the tail end was still on its way to its German targets.

All of our platoons were sending out patrols. Battalion was convinced that the only logical attack was to come off the sloping hills and silence the positions in the woods. The Ardennes forest was covered in a big, white blanket of snow. The evergreen trees found in this forest were much smaller than the trees of Northern California and Oregon that I was used to and they grew very close together, some in a straight line like they had been planted that way.

Our footwear was still pitiful. Some of us were still using the hole-filled black galoshes we had worn in the mud of France. We took a lesson from the dead Germans and lined the soles of these marginal GI-issue boots with soft straw for insulation.

One morning, after eating a hot meal of powdered eggs, fried Spam, coffee and hard biscuits and jam, I asked Captain Silver if I could speak with him alone. He had set up a bunk, table, telephone and radio in a small room, off from where the Belgians kept their livestock. Silver was smiling and still had a bandage over his forehead. He was wearing the overseas cap with captain's bars attached. I made some remark about the cap and his reply was that the battalion Colonel disapproved of it and had told him to get a helmet on. I then told the Captain that there was a man in the headquarters squad I wished to be moved out. He wasn't pulling his weight digging into positions and I knew he had gone to sleep when he was on watch. Silver looked me straight in the eye and asked if it was because the man was Jewish. My reply was, "No, sir!" The Captain knew better than that. I would put up with any man, no matter what color or pedigree, as long as he pulled his weight. All I was asking was for the Captain to find another place in the company for him.

I never realized until then how touchy Silver was of his Jewish background. I had heard slander towards Jews, blacks and even Italians (my own race) in the ranks. Silver knew how the master race (Hitler) was treating his own people, possibly his relatives. He had a purpose and he was performing above and beyond the expected. I heard him answer someone once, "Yes, I'm Hebrew, Jewish, that doesn't make me any more or any less than anyone else." Down deep inside, I believe his heritage was his purpose of serving his country. We all had a purpose driving us from deep inside. For a lot of us it was the pat on the back from a father, mother, brother or sister or maybe a big kiss and hug from a wife or girlfriend. We represented our families and, yes, our heritage each day on the battlefield. We would rather take death than failure. We often wondered what purpose was driving our enemy. He was just as cold, dirty, hungry and tired as we were, but the war had to go on. I guess he also did not want to return to his peers in Deutschland a failure.

Silver and I became better friends after that meeting. He called me "Babyface Giambroni" when the pressure wasn't on and told me he knew he could always count on me. In a couple of days the man I had complained about was moved from our squad to the kitchen crew. I believe he spent the rest of the war under the watchful eye of the mess sergeant.

Our kitchen crew should have been commended for how they followed us, the best that they could, and gave us hot chow in some very difficult times and weather. K rations were our main source of food. They came in small 12-inch weatherproof boxes, color coded according to what they contained. Breakfast was a small can with diced ham and eggs, a biscuit, prune bar, coffee, sugar, salt and pepper, a wafer and cigarettes. Dinner and supper were in different-colored boxes but they had one of the small cans of either cheese, Spam or meatloaf. They also had fruit and a candy bar and the other standards. Every box held four cigarettes. I got a small bottle of Kaopectate from the medic, but it had little or no effect on my diarrhea. The way we were living was giving us all a straight gut like a seagull.

Arloncourt was still a pain in the ass. We figured the best route was to take to the open hillside during the daylight hours and surprise the enemy by moving quickly down the hill. The half-tracks were to move in and out using their machine guns for cover power. The tanks and TDs were to follow with more infantry. This was the toughest assignment we had ever undertaken. To underscore how difficult it would be, two chaplains showed up that evening. We all knew that we would be up against our own artillery that would be firing in the woods for 15 to 20 minutes at jump-off time. We would have to elevate the fire power when we approached the woods.

My job was to fire the 50-caliber machine gun that was mounted on the turret of our half-track. The half-tracks were strung out over a one-half-mile area. Our track held just the driver, me and the communications sergeant with the bigger radio in back. The rest of the personnel were with the Captain and carried the mobile radio and battery pack. Years later I read that the German military's radio communications in the field was far inferior to ours. The larger radios we used were more active and our reception could get around or over the Ardennes mountains in most cases.

Company B was scattered over the knoll of the hill like a great big bunch of prairie lice. Our artillery was laying down a barrage of fire and I started to spray the woods with the 50-caliber machine gun. I had nearly run off a can of ammo when enemy artillery started coming in. Two GIs were hit next to our half-track. I jumped down from the track taking the aid kit with me and rolled over the first guy near the track. I recognized him to be one of the older guys in our company (30 or so). He was all torn up. The shell had hit directly in front of him and I didn't have to look twice to know he was dead. I scrambled over another 30 yards to the second GI. He was lying on his back and I told him to stay there and not to get up. He told me he was hit in the chest. I opened his heavy wool overcoat, pulled open his shirt and underwear and saw that a piece of shrapnel the size my palm was stuck in his pectoral muscle. I took my first aid packet out of my belt and removed the shrapnel with a fast jerk. One of his arteries had been hit and warm blood squirted me in the face. I pushed the packet, mostly gauze, into the wound to stop the hemorrhage, not even taking the time to put sulfa powder in the wound. Using a gauze bandage and his shirt I wrapped him up as tightly as I could and buttoned up his overcoat. The track returned and we loaded him in the back.

The communications sergeant told me not to stand up in the turret any longer and showed me the machine gun holes in the fatigues that I had draped over the turret. They were from a Kraut returning fire in response to me firing the 50-caliber. By now infantry and half-tracks were retreating all over the place. Some of our squad were already coming up to the track but the Captain was not with them. They said he was hollering over the radio for the platoons to pull back. I could see two half-tracks on fire. Most of the tracks had pulled back from the brim of the hill. Enemy direct fire and machine guns were causing us many casualties. Moments later the rest of our squad including the radio and Captain came back over the brim of the hill. I asked the men where Lejoy, a member of our squad, was. The last he was seen, they reported, was near the bottom of the hill. I ran over to the next half-track that had made it over the brim. Many in this squad were tagging along the side of the track. They said there was too much direct fire. Half-tracks can't stay very long on the brim of a hill because they would be too easily seen. I asked these guys if they had seen anyone from the headquarters squad. They too said no but that there was one guy they could hear hollering from a little draw to the left about 100 yards down the hill.

The Krauts had chased them up there with a machine gun and were firing from the woods. It was doubtful, they said, that the enemy would let me go down there. I asked the platoon sergeant if he could bounce in and out and cover me with the water-cooled 30-caliber machine gun on the track. He had three riflemen crawl out as spotters and, as he eased the track over the hill, I went down in 8–10 inches of snow. I ran about 100 yards down where I could hear him hollering from the small draw behind an old wire fence. It wasn't Lejoy, however, it was one of the new replacements. He had on his gas mask and a pair of leggings. He also had his bayonet—veteran soldiers had left all of this equipment in France. Both of his legs, below the knee, were mashed and held together only by what was left of those leggings. Grabbing him under the arms and trying to be as careful as possible, I pulled him through a hole in the fence. He was yelling at me because I had gotten rid of his gas mask and bayonet. He was a small man, weighing between 130 to 140 pounds, but the weight meant a lot going uphill, in the snow, with the enemy all around. Carrying him under the armpits, I ran uphill for about 15 to 20 yards and collapsed, barely breathing. The snow was kicking up around me from Kraut machine gun fire. I was pleased to hear our own 30-caliber returning fire and this gave me the confidence to get up again and drag him another 20 yards. Now I was only 50 yards or so from the brim. Three guys, I believe from the 1st Platoon, came down and two helped the wounded guy up to the half-track while the other guy helped me back up to the brim. I knelt there in the snow with the dry heaves from fright and exhaustion. I never did find out if that little guy lost his legs or even his life. The GI with the shrapnel in his chest, though, looked me up in Frankfurt after the war to thank me and show me the scar the shrapnel had left.

We had had the crap kicked out of us. We had lost men, equipment and at least three half-tracks. Some of the replacements we lost had only been with us a couple of days. Some of them had even been in the States for Christmas. Our company had been beaten up for over three weeks and morale was at an all-time low. The weather was still very cold and, while air power was giving us all the help they could, the Krauts were getting to be harder to find. They were masters of camouflage, especially in the blankets of snow where they used white overalls and helmets.

Chapter Ten
THE COLD ARDENNES

A couple of days after this battle we received more new replacements, equipment and supplies. We did not, however, get white coveralls or camouflage. General Patton was heard to have said, "My troops only need replacements, fuel [gasoline] and ammo and we will kick the shit out of those Krauts." Most of us hadn't had a bath since we left France. A lot of us would put on a second set of long underwear and call it good. I knew of one guy who had on three sets of long underwear. A lot of men wore black hairy beards, like Sergeant Mule. I could hardly even grow a fuzzy mustache that was why the Captain called Sergeant Taffy and me his babyfaced sergeants.

We had all lost a lot of body weight and were still losing the battle against diarrhea. We still had good appetites, though. I know when I could I would dig into our captured Germans' rations of canned liverwurst, blood sausage, head cheese and black stale bread. We also stole some 10-and-one canned rations from our own tankers. Ten-and-one, I never did ask what that meant. One GI said it meant that in one minute he could steal enough chow from the tankers to feed 10 men. The canned beans were my favorite from the cases we stole from the tankers. Frostbite and trench foot were taking a big toll on our personnel. We had to work to keep our feet and hands as dry and warm as possible. We would bed down in a manure pile or barn straw when it was time to sleep. You didn't smell so good, but it was much warmer.

We received good news one day: the Krauts had been driven out of Arloncourt by another battalion of the 6th Armored and the 35th Infantry Division. I heard that the TDs with the 90mm cannons were a better match for the Germans. Now we had to search for the enemy in the dark, cold forests of the Ardennes. The snow there was fluffy and we sank about six inches with every step we took. We sent out several scouting patrols. Our armored vehicles, including tanks, were of little or no use in the deep Ardennes forest. The mountains were not high, but it seemed like we were always walking uphill. We might not see the half-tracks at night, so we packed a heavy overcoat and wool blanket with us at all times.

A 1st Platoon scout reported that his patrol could hear male voices on top of a ridge over a mile away. To get to the enemy was to make a long, gradual climb to that ridge. The troops were as quiet as I had ever heard them. Our weapons were at the ready and my heart was thumping in my throat. Sergeant Ault of the 1st Platoon reported to the Captain that two enemy riflemen were just ahead at an outpost. The Captain ordered up all the automatic weapons we had, along with Ault and one of his men. Sgt. Ault and his man nailed both of the outpost Krauts who were wearing white uniforms. These shots brought enemy fire from the brim of the hill. I was with the radio and the Captain and saw the whole production get started. I really believed we were holding our own even with them having the high ground. Then came their mortars. The enemy were artists with a mortar tube. Their training with this weapon must have been intense.

The Captain said over the radio, "Shag ass—let's get off this hill." Soon all 125 to 130 men were in full retreat. We were lucky to come out with only two or three walking wounded. We stopped at the first clearing to catch our breath and to reorganize our troops. This clearing was a snow-covered road that carried through the forest. The Captain was standing only 10 or so yards from me and had just given the order over the radio to space out and move down this road. Sergeant Wooten of the 3rd Platoon was standing just an arm's length from me when the Captain told the 3rd Platoon to lead off. As Wooten turned, I asked, "How's it going, Woot?" His last word was "Rough," before an incoming mortar shell hit right in front of him. The explosion knocked all of us down. I reached for Wooten's boot in front of me and hollered, "Wooten, are you okay?" Several other mortar shells exploded around us. I quickly got to my feet and rolled Wooten over. He had taken the biggest percentage of the mortar shell and it was not a pretty picture.

Our squad was scrambling up the road carrying the Captain. He had been knocked unconscious from the impact of the shell. They sat him down a short way into the forest. He had come to, but was very groggy. The Exec officer came up with the Captain and conducted us into the woods using the snow-covered road as a guideline. When we cut back on the road again we came to a vacant house. The inhabitants had been gone probably two or three months. There was still some furniture in the house, but nothing else. Many of the European farmhouses were built fairly large with the barn and animal quarters built connecting to the farmer's living quarters. We took over the living quarters and barn area. It was getting dark and we all needed a rest. From our map it looked as though we were in Luxembourg.

The Captain had gathered his wits together and instructed the Exec office to tell the men not to smoke, no unnecessary talking, and to limit movement in and around the house and barn. We all took turns with guard watch and sleeping. It was snowing lightly outside. We hoped an enemy patrol would be discouraged by the weather.

It was in situations like these that our thoughts turned to home and family. I would have traded my place in heaven for a hot cup of coffee. I have had long cold nights since then, but nothing to compare to that January night in the Ardennes.

When morning came, the Exec officer contacted A Company and the 35th Infantry was moving through. It seems the Krauts could not find us in our hiding place. We concluded that the enemy was probably afraid we were going to sneak around to their rear and so they pulled back to a better defensive position. Silver had just pulled a maneuver that would have disqualified him in war games. The German-born military prima donnas would have said, "Dumb Americans, they don't even know how to fight with honest war tactics." We were still alive, but there were some things you would not want to try again. After a long march back to our half-tracks that were on one of the few roads still open in this part of Luxembourg, we had hot chow, hot coffee and all the cigarettes we wanted.

We had three or four slightly wounded from the last encounter. They were taken with the Captain to the aid station. Sergeant Wooten was the only man we lost. I can still see Wooten's torn body when I turned him over. My threshold for facing horrible sights like this was wearing thin. You felt like you had lost a member of your own family. I even got to cussing war out loud. It made no sense to give a life away for nothing, just to some maniac for power. Throughout the war I would say some prayers at night for Wooten, little Al and the many great soldiers of Company B who had given their lives.

Our battalion was in reserve and we got a great morale lifter—mail from home. The care packages didn't last long with GI knives always at the ready to help cut them open. Christmas fruitcakes were still popular among the bearded bastards of Company B. You didn't have to live long in our company before you became family. In the back of our minds we were always thinking that these guys might be the last brothers we ever saw. We also wondered if the American people understood how rotten war really was. This was supposed to be the war to end all wars. We know now that statement was not true.

This day we were the reserve battalion where we would get three hot meals a day and a chance to fix a warm place to sleep. We were camped in the cold snow and our squad had built three dugouts and lined the tops with fir tree logs. You could fit four guys in one of these dugouts and not be cramped. With the bitter cold, we really appreciated the wool blankets and overcoats. We took turns sleeping in our dugout, which was warmer than I expected. I really believe the cold Ardennes forest was warming up.

Forty to 50 prisoners marched through our camp and down the snow-covered road that day. One of the guards, of the 35th Infantry I believe, said that the German prisoners had just given up and had even called the Americans comrades. Ha. Ha. They looked like they were really tired of fighting. We stayed at this camp the last few days of January. We were not far from the Our River of the famous Siegfried Line. We could hear the sound of artillery to the east of us.

On the first of February we received the orders to pack up and move out. We left our little dugouts and all the hard work we had spent digging and building them. Our track driver gave me my first taste of schnapps, a clear liquid made from potatoes. He said it was called white lightning. It tasted like liquid dynamite. I asked him where he had got the juice. He said a civilian had given it to him in a little town as he passed through with the other tracks to pick us up.

Getting fresh water was a problem. We would receive several five-gallon water cans at a time but they were still arriving frozen solid. We needed water for our half-track and luckily a tanker gave us a can that had been thawed out. It seemed you didn't have to look for a problem—one was always on its way. The day we hit our destination, the sky had cleared up and we knew winter was on its way out.

Today, records show that the Battle of the Bulge comprised approximately 5 or 6 weeks and produced twenty thousand dead American soldiers, second only to the Civil War battles. All of the combined American divisions suffered 77 thousand casualties. The enemy suffered much higher losses in the Bulge, and it has been said that these battles were the turning point of the European war. Like one GI said, "If this is winning the war, I sure as hell would hate to be losing the conflict."

Chapter Eleven
THE OUR RIVER

We were on the edge of a small village, still in the country of Luxembourg. Our company had taken over some broken buildings on the steep banks of the Our River. Across the canyon and the river lay Deutschland, known in English as Germany. This part of Germany was called the Eifiel area and it had many rolling hills divided by small streams and rivers.

Things were pretty quiet in this area. The snow had melted and we had even seen some sunshine. Some of us bathed out of our helmets. I couldn't remember my last real bath–I think it was in France the previous fall. They brought in a mobile shower trailer with both cold and hot water trucks attached. In the shower tents, a group had two minutes to soap up and two minutes to rinse off. The system was like a triangle and you had to be quick. You laid your clothes on a bench in the first tent, the second tent was the shower trailer, and in the third tent you were given a bath towel. You dried off there and received fresh underwear and socks, then returned to the benches to get dressed.

The Exec officer had me look over the topographic maps showing the west banks of the Our River. He wanted a couple of machine guns spotted in an FPL line (Final Protection Line). This was to provide the greatest cover from the approaching area. We were planning away when out of the blue someone in our crew hollered, "Attention!" I continued looking out of the broken side of the building matching the map area. It was suddenly so quiet that I turned and stared into the two stars on the helmet of Major General Robert W. Grow, commander of the 6th Armored Division. I showed him what we were doing and he seemed to be impressed. He then asked me where our company commander was and I said that the Captain was back at the hospital and the Exec officer was at the CP in the next building down the road. The General and the officers with him left. He seemed to be a very pleasant man. I'm glad it wasn't General Patton; they say he would have expected my boots to be shining. I was really glad, though, that we had had time to clean up a little before General Grow showed up.

After the first week of February, our outfit crossed the Our River on a pontoon bridge built by the combat engineers. On the east side of the river was the famous Siegfried Line that strung from Holland to Switzerland. This area had pillboxes, anti-tank trenches and pillars to protect Deutschland from any aggression from the west. The 6th Armored found many ways to penetrate this line, though, and in a matter of hours these enemy positions gave up. Over the next two weeks, I learned the 6th Armored took more than one thousand prisoners. Two or three villages were falling to us in a day. German equipment and weapons were left in the fields when the enemy garrisons would pull back.

It was nice living in the village homes that had not been damaged. Unfortunately we weren't there for long; the armored division kept in high gear. In one village, one of our platoons captured a horse and wagon that was carrying soup, milk and dark bread for the enemy troops. The two German soldiers in the wagon, one riding the horse and the other

February 1945—Baily Bridge crossing the Our River,
on the border of Luxembourg and Germany

in a seat, were surprised the American troops were already there and that their comrades had left before dinner.

There were places where the German troops would set up their machine guns and fire mortars before retreating. I could tell the sound of a Kraut machine gun and the color of their tracers; they were different from the GI equipment. These machine guns were much faster and the tracers looked purple, especially at night. Their artillery was still causing many casualties. A weapon the Krauts started using in the Ardennes was now being used more often. This was what us GIs called the screaming Mimi. It had a series of six tubes that fired rocket-like projectiles. They made a whining, terrible noise in flight, and created multiple explosions when they came down. The first time I heard it fired it made my pucker string twitch. Lucky for us, the enemy was never very accurate with this weapon.

It was at the Our River where I received a large package from California. It contained enormous amounts of Italian foods like salami, cured ham, pan dulce (sweet bread), cookies and candies. This came from an Italian club in my home town where my mother was a member. The contents were divided democratically among the members of our squad. I got two salamis that I carried in my pack and snacked on. There also were several pairs of wool socks in the package. I got one pair but some guys only got one sock. I said one sock was better than a sharp stick in the ass. We did smile and laugh a little every now and then. Humor always seemed to release the pain of war.

The pillboxes and tank traps in this area extended a few miles to the east. They were manned by six to 12 occupants. The tank's 75- and 90mm ammo would just bounce off these cement steel blocks. They were sturdy enough to even endure a direct hit by a 105 howitzer. The TOT (time on target) that the artillery and FO (forward observer) put on single pillboxes did, however, get results. Many white flags came out of the turrets of these boxes. TOT was a configuration of many guns (cannons) on a single target. I heard that the combat engineers also used dynamite charges on the rear entrances of the pillboxes. They said smoke billowed out followed by bunches of staggering Krauts trying to hold their hands up in surrender.

Enemy artillery was still causing us casualties. They would fire their automatic weapons as long as they could, especially at night, then they

*Leon Digby, a member of the headquarters
squad, from Anniston, Alabama.*

*Part of the headquarters squad. Rear, left to right: Vargas,
Andy Giambroni, unidentified soldier (remember the face, but not the name).
Front, left to right: Communications Sergeant "Pappy" and "Red" Bryant.*

would flee their positions leaving their wounded and many of their heavy weapons, mortars, anti-tank guns and, in one incident, a light machine gun behind. The enemy was headed for the Rhine River. This they knew, and they figured they could defend it and keep Deutschland free for the Fuehrer. The Rhine is a very large river, comparable to our Mississippi River for the most part. The Our River was very small compared to the Rhine.

At this time of year, all of the rivers and streams were flowing fast from the melting snow. The snow was almost completely gone and the weather was much warmer. The combat engineers were kept very busy checking for mine fields. The Krauts were using every method they knew to delay our advancement. Dead horses lined the roads of their retreat. These horses were used to pull their artillery pieces. We even found some artillery pieces left behind with some of the dead horses. "Nix Benzine" (no gasoline), the German military would say. Their Air Force was of little use without fuel. Most enemy planes we sighted were being used for observation only.

Units of the 6th Armored were said to have been the first Americans to invade Germany. There was a sign posted in front of a pillbox on the east banks of the Our River in Dasburg, Germany, that read: "You are now entering Germany through the courtesy of the US 6th Armored Division."

Several of the soldiers who were in the headquarters squad came from the southern states of the US. These were brilliant young men who never went to school or may never have had the opportunity, especially during the depression years of the thirties. It was hard to believe that there were so many soldiers without formal education. This was in the 1940s and many could not read or write. I would read letters to them and sometimes I would even write their letters. There always will be a warm spot in my heart for these guys.

The guys were masters of improvisation. They took cowbells they had found in a barn and a ball of string and fastened them together to create an alarm at the avenue of approach to our dugout. The only guy to ring the bells was a BAR (Browning automatic rifle) man, looking for the 2nd Platoon. Regardless, it gave all of us a sense of security while lying down in our dugout. These guys could dig out two or three covered dugouts in short order. Leon said he felt like a ground squirrel—every time we stopped he wanted to dig a hole.

The war had changed by this time. The German Army was leaving the towns and villages and was fighting delayed-action maneuvers. They began fighting with more artillery, and anti-tank and anti-personnel mines. They destroyed electrical power, bridges and main travel roads as they pulled out on their way to the Rhine. They figured this large river would stop the Allied Forces.

The combat engineers of the 6th Armored cannot be praised enough for getting our armor across the small streams and roads and clearing the increasing number of land mines. The fields and roads were littered with numerous land mines. The enemy had put out hundreds of mines, both anti-tank and anti-personnel. Unfortunately, the combat engineers did not find every one. We suffered our share of losses to the anti-personnel mines. From what I understand, Company A was hit the hardest out of our battalion.

The anti-personnel mines were triggered so that a person walking could set them off easily. They would kill the person sometimes, but mostly just blow off one or both legs. This is a very cruel type of weapon and one that was feared greatly by the infantry. Even in the delaying action of the enemy, we had many casualties. Some of them died on the way to the aid station, but most of those we could get there survived. Our medical support group did an outstanding job with not a lot to work with. The aid station tried to keep as close to our battalion as possible. Most of the time they were within artillery range. The artillery shells didn't read the red crosses on aid station tents or the buildings they used.

Chapter Twelve
LUNENBACH

February was winding down and our company found itself on the high banks of the Prün River. This river was not too large, about 50 or 60 feet wide, but was very full from the snowmelt. I had the radio and was with Captain Silver, who had returned from the hospital. We had taken shelter from enemy artillery under a railroad trestle. Our objective, the town of Lunenbach, was just across the Prün River.

Captain Silver said we couldn't take the chance on taking a tank or TD across the river with equipment. We had reels of wire, telephones, ammo and heavy weapons that were to help us once we got to the other side. Leon said he had seen a span of horses in a barn above. In addition to the horses, there was a wagon and a harness for the horses in the barn. I had never heard the term "span" before, but a span of horses is two abreast. The Captain asked if I thought Leon could take a wagon and our gear across the river. "Captain," I said, "he will go across or die trying."

Leon hooked up the horses, and everything heavy such as the wire reels, telephones, radios and weapons was loaded on the wagon. We were ready to go down the steep hill to the river bank. We planned not to enter the river until it started turning dark. At dusk Leon put a long pole between the rear wheels of the wagon. Leon called it rough-locking the hind wheels so the two horses could hold back the wagon and slide it to the river bank. We waded with the wagon across the cold, waist-deep Prün River while Leon rode in the wagon guiding the horses. We were in the middle of the river when one of the horses slipped and fell. Leon jumped off the wagon into the cold water still holding the reins of the horses. This southern boy got the downed horse up and he walked and drove the wagon to the bank of Lunenbach.

Company B scattered throughout the town. The enemy had retreated but now German artillery was shelling Lunenbach. Silver gave orders where each platoon would set up defensive positions. He had assigned our squad to get all civilians DPs (displaced peoples) into one common cellar. (See Before and After photos of Lunenbach.)

We had just finished the ordeal with crossing the Prün and I was ordered to take the headquarters squad through the small town and to gather the civilians into a cellar around the center of town. This was not an easy task since the Krauts had destroyed electrical power and incoming artillery was starting to increase. The squad went in pairs in all directions. One pair from the squad came to a cellar with some DPs but said that there was a young woman trying to have a baby and that they had left her and her parents where they were. I got a man from the 4th Platoon who spoke German and we went to investigate.

She had been in labor a couple of days and her parents asked us to please get a doctor. I called on the radio to the aid station but their hands were full treating the wounded and they had no way to cross the river. Their advice was to do the best we could. Strause was the GI from the 4th Platoon. He sure was valuable in translating German for us. We went to another cellar and saw one woman with children around her. I had Strause ask her how many children she had and she said "*Drei*"–three in English. I had him ask her to come with us–she was three times smarter than me in delivering babies. My total sum of experience in obstetrics was watching a childbirth film and helping my veterinarian brother deliver a calf.

In a bedroom in the upper story of the blacked-out building was the young German girl, 20 years old or so. She was grabbing the rungs of an old iron bed with both hands to help her deal with the pain. This was her first child. We all had wet cloths but I made sure that we all washed our hands. My shirt was off and I had scrubbed my arms and hands and cleaned under my fingernails. Our crew was made up of two guys from our squad, our medic (a truck driver in civilian life), Strause, me and the midwife. The girl's parents stayed in the kitchen and boiled up a pair of bandage scissors and a roll of bisect tape. Heavy clean sheets and towels were put under the girl's lower body and legs. The midwife palpated the lower abdomen of the young girl and looked at me and said, "*Kopf*," or head, meaning the head of the baby. She then pressed down as she was turning the head from the the outside of the body towards the birth canal. This increased the pain so I asked the medic if he had anything for severe pain. He had some blue 88s, short for Phenobarbital. They say you can give morphine to a patient if they're not shot in the head. She qualified for the latter but I only gave her half of the syrette, which amounts to about a quarter grain of morphine. Within twenty minutes or so she was relaxed. The midwife was

Map of the European war, March, 1945.
The 6th Armored Division was just south of Prün, Germany.

now pushing on the head again; the morphine had done the trick. She was more relaxed but still conscious and aware of all the things the midwife was saying. Looking into the birth canal you could see the hair on the head of the baby. There was a big smile on the face of the midwife. Within a few minutes the young mother delivered her baby, a boy. I tied the umbilical cord in two places, one near the baby and one leading to the afterbirth (placenta). The job was okay, but I think I made this little boy's belly button longer than any in Germany. The midwife had two pans of water, one lukewarm and the other cold. She dipped the baby back and forth, washing its little body. I saw the little guy gasping for air before he ever made a sound. The baby was wrapped in blankets and then the midwife tended to the girl and the afterbirth.

It's strange the things you remember. There was so much urine expelled from the mother after the birth of the baby, and the odor stayed in my hands no matter how hard I washed them. The parents put their arms around me and the guys. The father had dug out a bottle of schnapps which I took one shot of on an empty stomach. I knew better than to indulge any more. The rest of the crew stayed drinking schnapps and frying eggs. I went up to the CP where the gang was eating hot chow. The mess sergeant had gotten a TD to come across the river, no problem. I took some razzing about bringing a baby Kraut into the world. Silver tried to talk me into eating, but all I had was some pear juice and a couple canned pear halves.

About the second day of March we got on our way again. The enemy had retreated to new positions but they were still dealing out artillery. Our P-47s were discouraging a great deal of the retreating artillery. The enemy had put out a large series of mine fields including both anti-personnel and anti-tank mines. Sad to say, I heard that Company A had run into a large mine field. The anti-tank mines took a vehicle to trigger them. The anti-personnel mines were to cripple the foot soldiers. They were easy to set and many could be put out in a short time. There were two anti-personnel mines that I was acquainted with, the Bouncing Betty and the Shoe Mine. The Bouncing Betty would explode once stepped on and fly in an arc for a second explosion. The Shoe Mine was a simple small cigar box with the lid a fraction of an inch open. When it was stepped on and the lid closed, it was ignited and a powerful explosion would take a man's foot or leg off.

On this day our radios needed new batteries so the Captain told me to take two men and go back to the CP to get a set. Taking his orders, we went back over the rolling hills to the CP and got the batteries. At the CP, the 1st Sergeant told us that the Jeep driver would take us back to the company. It was a mile or so across the hills, but it was more than two miles around the road. We were going to ride in the Jeep so we were not in that big of a hurry. We were about five minutes down the road when we saw a Sherman tank that had been knocked out by a large anti-tank mine. I said the ground was okay so we pulled out around the tank. We had not gone more than 50 feet off the road when the left front wheel of the Jeep was blown off. I told everyone to sit still, we had hit a Shoe Mine. I instructed the two guys in the back to go over the hind wheel tracks to the road and not to vary one inch either way. The Jeep driver and I followed them and we returned to the CP. We went back to the company afoot the way we came. The anti-personnel mines were terrifying. They kept us on our toes, eyeballing the ground everywhere we went.

The days had started getting longer; the snow was gone but it was still very cold at night. I had been given an introductory book of veterinary science from the English USO and I packed this book in my pack with my leather binder and writing supplies. Our squad would have me read aloud some of the chapters of this book to break the long hours of waiting. They would love for me to recite big words like equine encephalomyelitis, known as sleeping sickness in horses. I really was interested in animals. I had hopes of being a veterinarian and owning a cattle ranch. Delivering that baby confirmed that being a human doctor wasn't for me.

It was late in the morning of March 3rd when B Company moved out on foot. The half-tracks were left back and we strung out along the road for a mile or so. The headquarters squad had the radios with the Exec officer. The enemy shelling was becoming more intense and they were using a lot of the Screaming Miemies. The shelling was so close that many times we had to hit the dirt along the road. A shell hit about 20 feet in front and to the right of me. Shrapnel hit my helmet, the gun stock of my carbine and the end of my terminal finger on my left hand. I pulled my glove off and the finger was bleeding like a stuck hog. The shrapnel had torn open the mitten I had on and opened the end like an oozing sausage. I went to the medic who was not far behind me and he rolled a bundle of gauze to stop the hemorrhage. He did this while we were both halfway lying on the ground.

The Krauts didn't slow up their shelling one bit. In front of us we heard someone calling for the medic. He went forward and I said so long to my squad and half walked, half ran back to our CP. Silver and the 1st Sergeant were there as well as a Jeep driver who took me to the aid station, a short distance away in the torn-up village of Lunenbach. The doctors there redressed the finger and said that I would need surgery because part of the bone at the end of my finger was gone. I was smiling all over. This was truly a million-dollar wound. I was going to the rear after six months of living like an animal and praying to stay alive. I asked the medical doctor if he had seen the girl who had the baby in the town. They had, but informed me that the baby didn't make it. The mother had been in labor for too long and her system was too uremic. I was told, however, that our work had helped save the young mother and that she was still in the same house and doing fine.

Note: I returned to Lunenbach with my family 53 years later and tried to find the woman. We were told that an older woman who may have been her had died a few weeks before we arrived. The people we met were going to write me to tell me her name and if this was the same person. More than two years have gone by and I haven't heard a thing. I will make one more trip to Europe, I hope, before I die. I would at least like to know her name. Some of the people of Lunenbach said this woman was probably not German but from one of the agricultural laboring families from Poland or other countries that the Nazi regime had brought in. This part of Germany is primarily agricultural and the same German families have lived there for many, many years. The scars of war are gone from the area now. I mean they are gone from the farms, roads, autobahns and buildings of beauty. The only scars left are the scars in the hearts of the people that lived through that war. When I meet older people, 70 years or older, the war is never discussed by them. They will answer questions about the war but it will end there. Note the pictures of 1942 and the damage done in March of 1945 to Lunenbach. The mayor of the town had both of these pictures. I told him that the majority of that damage was done by German artillery. The German Army was headed for the Rhine River and they destroyed anything that would help the Allies.

During this visit I asked the mayor about the railroad across the Prün River and he said there was no such railroad. We drove across the river to the bank where our Company B crossed 53 years ago. The German owner of the property showed my sons where the railroad went. He had purchased the railroad property about 20 years after the war.

Lunenbach, Germany, before World War II.

Late February 1945—The city of Lunenbach, Germany, after two to three days of fighting. Most of this damage was done by German artillery and Air Force P47 bombings.

Early March 1945—Lunenbach, Germany, a few days after fighting ceased. Church steeple remains, even after heavy bombing.

July 1998–Lunenbach, Germany, as it looks today.
Note that the church steeple is still standing.

*As we start through Germany,
lines of prisoners were abundant.*

*Second Platoon track. Note the soldier (left rear)
wearing a German officer's cap.*

Looking back at our column and another captured German village.

Chapter Thirteen
MILLION-DOLLAR WOUND

The very evening I was wounded, my arm in a sling, I climbed into an Army ambulance with a wounded German soldier on a stretcher. The German had a badly shot-up thigh and lower leg. I felt guilty with my minor million-dollar wound—maybe he hadn't prayed as hard as I had. Casualties must have subsided because we had the whole ambulance to ourselves. I lit up a cigarette and took a couple of puffs when the enemy soldier said, "*Zigarette, bitte*" (cigarette, please). I looked at him and thought to myself, a couple of days ago this guy would have blown my head off, now he wants to be friends and enjoy a smoke together. What the hell, I never could figure this war out anyway.

I put the cigarette in his mouth and gave him a light. He said "*Danke*" about three times—thank you very much. My German wasn't good enough to strike up a conversation but we both smoked about three cigarettes each before we got to the evacuation hospital. Where this evacuation hospital was, I never asked. It was a series of tents. The German went to the prisoner ward and I was put in the ambulatory ward, or the ward for the walking wounded.

The next day I was taken into the surgical ward which was in side tents off a long hallway. The doctor who put my middle finger together had graduated from Stanford University. Stanford was in Northern California, not far from where I was raised. Across the aisle from me surgery was being performed on a GI. I think they were amputating his leg. What a bunch of bullshit war is, I thought. Sometimes I wonder if politicians, big business or even the average American really understood the horrors of war. Only a small percentage of our armed forces ever got into the front lines of the fighting.

I slept good in a warm cot that night. For the next few days us guys in the ambulatory ward assisted in feeding wounded soldiers in other wards. A nurse had me feeding a badly wounded black GI. He was bandaged and had casts over most of his body, including both of his arms and legs. I understood him to say he was in a TD platoon. He still had a lot of humor in him, with such a beat-up body. He said, "You know I was a spearheader for that man Patton, but thank the lord I'm going home." He would chuckle and smile under all those bandages. In another ward I recognized a platoon sergeant of A Company, my same battalion. The sergeant and several of his platoon were in that ward. They had lost their lower legs to Shoe Mines. The enemy had put out several of these mine fields. I was only there three days altogether before several of us walking wounded were loaded into an ambulance headed for a general ambulatory hospital in Bar-le-Duc, France.

The Bar-le-Duc hospital was set up in a convent. It had been a school at one time and the nuns had lived next door. The convent had large wards. There was a wall around the convent and a fairly large courtyard. The first night I was in the hospital, March of 1945, I was given a new medical drug, penicillin, in the muscle of my butt every four hours. A medic came up to my bunk in the middle of the night, tapped me on the shoulder and jumped back. I had one eye open and asked him what the problem was. He told me it was time for a shot but that guys like me who were fresh off the front were a little hyper. I said, "Soldier, you can stick me in the ass with a bayonet, you're not going to get me out of these clean sheets."

In the month that I spent at this hospital I gained almost 20 pounds. We were able to get passes almost every afternoon as long as we were back by bed check. Money was no problem at that time—in Bar-le-Duc we sold soap, cigarettes, cigars, gum or candy.

I was healing up real fast. The Captain who was the commander of the hospital had a formation by our beds each morning. He told me I was ready to go back, but if I was to take the other ambulatory patients and march them out each morning and back by noon each day (except Sunday, that is) he'd keep me a little longer. He told us to remember that we knew how to march like soldiers. One half hour after breakfast each morning we would fall out into the courtyard and line up in columns of two. All these men had been wounded but needed exercise to help their healing process.

The first morning after I had taken roll, I said to these soldiers, "The Captain said if I can march you out and back into this place like soldiers

I can hang around a little longer. I don't know about you, but I don't want to go back too soon. We will march out here until we get out of sight—I know a place we can take a break, not far from here. Then we will start marching back when we see the hospital and we won't be late for lunch." From the nods and smiles I got I knew they were with me.

The American GI fascinates me on how fast he can catch on. We had a place to take a break near a small restaurant's beer and liquor garden. The men seemed to like beer more than cognac and other liquors. They took care of themselves and we would return to the courtyard marching and counting cadence. The French girls, mademoiselles, liked to visit the American GIs while we were resting, snacking and having a drink. They were very pretty, wholesome and would bring fresh sourdough French bread and cheese. Most of them could speak some English and it was a treat to listen to them speak English with a French accent. It was no problem to get a formation and march out of the gate each morning. It was only on a few rainy days that we did not go out. We stayed inside at our beds, writing letters and reading books.

One of the things I thought about while in the relaxed atmosphere of the hospital was whether I would ever get married. The girl back home was not even in high school. She was just a little kid then, the girl I would eventually marry. I had to get back into college and become a doctor of veterinary medicine, I thought. To become a doctor would take at least five or six years. That summer I would be 22 years old. The war was looking like it would end soon, but still it would take a year to get home and back in school. That would be 23 and five so I would be at least 28 years old by the time I could become a doctor and afford to get married.

I was anxious for this war to end so people, including myself, could go back to normal lives. I wrote a lot of letters home—I was really homesick. I also wrote letters to Colorado State University Veterinary School. My home state, California, did not have a veterinarian school at that time. Colorado State's Dean Newsome of the vet school wrote and told me that I could apply when I was discharged, but that I had to have all my credits before applying. He also said that the classes were only big enough for 60 students.

I remember the day we heard President Franklin Delano Roosevelt, our Commander-in-Chief, had died. We had a special formation in the courtyard and the flag was raised at half-mast. Harry Truman was sworn in as the President of the United States. It was the middle of April. In a few days my name was posted to report to a repo depot replacement center in France.

I hated to leave the hospital, but all good things have to come to an end. The medical officers, nurses and other medical personnel had been extra great to all the GIs who were there. They had shifts around the clock. Most of the cases handled there were sent back to their units healed up. Cases that needed special treatment were sent to the hospitals in Paris and England. The real tough cases, amputees and very serious wounds or conditions, never went there. They were treated in other hospitals and sent home to the states if they could stand the journey.

This hospital did handle some cases of battle fatigue. These poor guys had their nervous systems torn up. In World War I they had called it shell shock. The GIs suffering from battle fatigue had a lower threshold for the noise and brutality of war than most. Many of these GIs never returned to the front; instead they were used in non-combat jobs in the rear with supporting units. These men had been wounded, just not in the flesh and blood. I was lucky that my threshold was stronger, but I know it came very close to cracking sometimes.

Our medical hospitals were completely separate, other than the evacuation hospitals, from the wounded enemy. The next morning I was supposed to leave with another 10 GIs. I remember we played cards that evening but my mind wasn't on the card game. I felt I had been gone from our Company forever even though it was less than two months. It was the hurry-up-and-wait process of the Army all over again.

I wasn't too worried to be going back. By then the Allies had crossed the Rhine and the armored US columns were storming all across Germany. From the north and the south into Austria, thousands of German prisoners were being taken. They wanted to surrender to the Americans; they did not want to take a chance of being taken prisoner by the Russians.

I spent a lot of time trying to locate my division. The 6th Armored was with the 3rd Army of General Patton. It seemed like the 3rd Army was zigzagging across Germany. They had liberated a couple of the concentration camps that the Germans did not like to talk about. This was not an isolated case, it was happening all over Germany. It is hard to believe that human beings could torture other humans and be able to sleep at night. The German people turned their heads to these atrocities. Germany (the Nazi Regime) was coming out of the great depression. Whatever the Nazi tactics were, they were bringing economic relief to the German people. The German people sold out their freedom, such as it was, to the Nazis for economic reasons. After the war these were the excuses I got from the German people.

Chapter Fourteen
COLONEL JIM MONCRIEF

Having spent this time at the ambulatory hospital in Bar-le-Duc, I had missed the crossing of the Rhine and the fast advancement of the 6th Armored Division through Germany. I was fortunate enough to meet with a former division headquarters officer of the 6th Armored Division 53 years later at a Super Sixth Association meeting in Colorado. He is Colonel Jim Moncrief, personnel officer to General Robert Grow of the 6th Armored. How lucky I am to be able to collaborate with a man of his experience. He is 88 years of age at this writing and is real sharp. He is the author of two books, *As You Were, Soldier* (1996) and, *History of Co B Ninth Armored Infantry Battalion, Sixth Armored Division*. He has given me consent to use any parts of either of his books and, as I missed this time in combat, his recollection is what follows.

I have included Colonel Moncrief's biographical sketch at the end of this book. It shows his magnificent 30-year military career. You can tell he is a true infantryman from his article "Close Combat," also included in this book. He was one of the first officers to inspect the Nazi Buchenwald prison camp after its liberation by the 9th Armored Infantry Patrol of the 6th Armored Division. Colonel Moncrief was not only an officer in the division headquarters of General Grow, he was his personal friend until the General's death years after both were retired.

BUCHENWALD

By Colonel Jim Moncrief–As You Were, Soldier *(1996)*
Reprinted by permission of J. Moncrief.

Following the breakout of Bastogne in the ice and snow, the tough fighting through the Siegfried Line, the Rhine crossing at Oppenheim, and the capture of Frankfurt, the Sixth Armored Division began what many of the troops called "The Rat Race." The German Army by the first of

April 1945 was in a state of disarray. Faced with the rapidly advancing American forces, the Germans were retreating just as rapidly. They could not organize their fleeing forces to present a formidable defense. Operating under the XX Corps in Patton's Third Army, the Super Sixth followed a route north and east out of Frankfort towards Giessen, turning almost directly east just south of Kassel toward Eschwege and Mulhausen. Serious fighting was encountered at Mulhausen and Langenselza, before moving further eastward toward Zeitz and Bad Sulza.

The Division's southernmost column included a Task Force of the 9th Armored Infantry Battalion commanded by Capt. Robert J. Bennett. A part of this task force was a four-man patrol in a scout car vehicle commanded by Capt. Fred Keffer. (NOTE: Nobody has bothered to determine the origin or source of the scout car, an unauthorized piece of equipment for our division.) In a little village, Keffer and his men encountered several men in a squabble on the roadside. They were yelling and fighting with their fists.

Among the four men of Keffer's patrol was Sgt. Herb Gottschalk, who could speak Polish, German, Russian and perhaps as many as two other languages as well or better than he could speak English. Gottschalk was able to determine what these street fighters were saying in German and Russian. From the jabbering and yelling, Gottschalk found that some Russians had been in a nearby prison camp and they were chasing and fighting several Germans who had been guards in the camp. He further determined that when the Germans, upon hearing the advancing American troops, had abandoned the camp; the Russians who had been prisoners in the camp for only a short while, decided to give chase and catch the German guards. Also Gottschalk was told that the camp had thousands of civilian prisoners dying and in various stages of starvation.

Keffer radioed Bennett for permission to leave his assigned mission long enough to investigate the reported camp, which contained thousands of starving and dying civilian prisoners. Bennett approved the venture, cautioning Keffer to rejoin the column ASAP.

Putting a couple of the Russians on his scout car, Keffer was led to the site of the secluded camp which was only a short distance out of Weimar, one of Germany's principal cities. In an article published in *THE SUPER SIXER* in the fifties, Keffer described his reception. The

Americans were hugged, repeatedly thrown into the air, and cheered by the more able-bodied. Restoring order, Keffer assured the inmates that help would be there as soon as possible.

Thus it was the fate of the Super Sixth to discover Buchenwald, one of the Nazi's most infamous concentration camps. It was a camp of

James S. Moncrief Jr., Colonel (retired) US Army.

Courtesy of Colonel James S. Moncrief

horror where Hitler imprisoned Jews and political prisoners. Many thousands were killed, starved to death, or died from maltreatment. (Note: I have pictures which graphically depict some of the atrocities committed by the German SS guards at the camp).

At the time there were approximately 21,000 political, criminal, and war prisoners in the camp, many of whom were just barely living. Another 40,000 of the more able-bodied were reported to have been marched eastward four or five days earlier.

Keffer sent a radio message to Bennett and higher headquarters, which was relayed to Division Headquarters, briefly describing the terrible situation involving thousands of prisoners at the camp which he had encountered. When the message was received at Division, the staff members knew that some immediate action was required to get assistance and relief to the camp.

Fifty years later, I really don't remember whether I was ordered or whether I voluntarily went to Buchenwald to determine more detailed facts. Insuring that Corps and Army Headquarters were notified of the discovery of the camp and dire need of aid for "thousands" of dying and starving inmates, I went to the camp site by jeep.

As G-1 (Personnel Officer) of the Division Commander, I had staff responsibility to represent the General in matters dealing with people—including prisoners. I was there in about two to three hours after Capt. Keffer and his patrol had discovered the camp. I was accompanied by the jeep driver, whose name I do not remember. I do not recall whether or not other members of the Sixth were there at that time or not.

Reaching the camp, I saw the "yard" with a lot of inmates wandering around. A few of them looked to be in fair condition and were able to maneuver. But most of them were in terrible physical shape, barley able to move along. Some were only half clothed. They were very thin, a mere skeleton covered with skin. Most of them had a blank stare, with a hint of a pleasant, but toothless, smile, as I approached. I also saw many bodies laying on the ground. I remember thinking, at first, that these were men taking a nap. But I soon discovered that they were dead. I was directed to a building which was apparently the center of activity, or the headquarters of the inmates.

After fifty years, my memory will not permit me to replay all the details of my short visit there, but I found a young inmate who could

April 1945—Dead prisoners at a political prison camp.

April 1945—Crematorium at Buchenwald.

speak some English. He turned out to be a young Austrian Jew who had been a pre-med student when he was interned. The other inmates referred to him as "Docktor." His physical condition appeared to be much better than any others who I saw. I remember thinking at the time, that he perhaps had been able to get more food than the others.

My guide showed me through one barracks where I saw other bodies laying on the floor. A few men were standing around. There were several in their bunks, which were stacked wooden platforms or shelves. There were no mattresses or bed linen. Those who were in the bunks barely had room to turn over, the space between bunks being so small. Most of them were so nearly dead that a stranger walking through the barracks attracted very little attention. Many did not have enough strength remaining to roll their eyes to notice the stranger.

Word had been spread, however, that the Americans had come. Practically all of the "standing" victims had a "sort" of grin of appreciation as they nodded toward me. All who were awake and reasonably alert wanted to touch me.

I remember my guide showing me the Crematorium with its two huge ovens. He explained to me that the holes which were dug into the walls (beaverboard type) were made by those victims who were hanging by their feet waiting to be placed into the ovens.

By this time in the war, I had been quite accustomed to seeing death, dead soldiers and civilians. But I had never seen walking death. I had never seen so many dead at one location. I had never seen men in such numbers, all of whom appeared to be at death's door. I will never forget the expressionless stare in their eyes.

My purpose of going to the campsite was to obtain an eyewitness estimate of the situation, to communicate directly with the Division G-4 (Supply and Logistics Staff Officer) to have him expedite supplies of all sorts to relieve the suffering of the victims at Buchenwald.

The G-4, Caleb Boggs (later the Governor, then Senator from Delaware and now deceased), was able to obtain some supplies and medical attention from within the resources of the Sixth Armored Division. But more importantly, he was able to get assistance from the Third Army with its far more ample resources. When I returned to Headquarters, I remember being in his office that night and the two of us were in direct telephone communication or radio, (I don't really remember) with an officer of Third Army G-4 section about immediate needs: food, kitchen equipment and personnel, water purification, medical personnel and equipment, etc.

Since the mission of our division was to pursue the German Army which, by that date, was in some disarray and "on the run," our division could not tarry at Buchenwald. It remained for other units from Third Army to come in and administer to the victims.

Gen. Patton later ordered conducted tours of the camp for many of the leading citizens of nearby Weimar, when they professed total ignorance of the camp and its horrors.

A Survivor of Buchenwald

by Colonel Jim Moncrief—As You Were, Soldier *(1996)*
Reprinted by permission of J. Moncrief.

In the spring of 1945, the Sixth Armored Division overran Buchenwald, one of the Nazi's most infamous concentration camps. It was a camp of horror where Hitler imprisoned Jews and political prisoners. Many thousands were killed, starved to death, or died from maltreatment. (Note: I have pictures which graphically depict some of the atrocities committed by the German SS guards at the camp.)

A patrol under the command of Capt. Fred Keffer of the 9th Armored Infantry Battalion, a unit of the "Super Sixth" learned of the camp's exact location from some Russians who had managed to escape from the camp only a short time before. In a nearby town, the Russians were in the act of attacking former German guards of Buchenwald as the latter were fleeing the onrushing American forces. Among the soldiers in Keffer's patrol was Sgt. Gottschalk, who could speak both Russian and German as well as Polish and Italian. He was able to understand the problem between the former prisoners and former guards. At the point of American guns, the Russians were finally separated from the Germans.

The Russians then led Keffer and his patrol to the site of the secluded camp which was only a short distance out of Weimar, one of Germany's principal cities. At the time there were approximately 21,000 political, criminal, and war prisoners in the camp, many of whom were just barely living. Another 40,000 of the more able-bodied had been marched eastward four or five days earlier.

When the message of the discovery of the camp was received at Division Headquarters, I immediately went to Buchenwald to determine some facts so that the division could report the details to Army Headquarters. Gen. Patton later ordered conducted tours of the camp for many of the leading citizens of nearby Weimar, when they professed total ignorance of the camp and its horrors.

Forty years later, in May of 1984, I was a member of a group of about 70 former members of the Sixth Armored Division which had the pleasure of touring portions of Europe where we had fought. In Bastogne, the American Ambassador to Luxembourg, Honorable John Dolibois, and a distinguished looking citizen of that country joined our party. The white-haired old gentleman was introduced as Leon Bartimes, the Mayor of Beaufort, a small town near Clervaux, Luxembourg.

Parenthetically, a statue of an American GI soldier stands in the center of Clervaux, dedicated to the memory of the Americans who died to liberate that city from the Germans. The insignia of the Sixth Armored is prominently displayed on that monument.

Visiting Hamm, site of the American Cemetery where 157 former members of our division are buried, a Memorial Service was conducted at the gravesite of Gen. Patton. Mr. Bartimes joined in with all the "Super Sixers," doffing his newly-acquired white Sixth Armored cap at the Memorial Service with the others.

Following Hamm, the American Ambassador told us he and Mr. Bartimes wanted to take us to Beaufort. That rip was not on the original planned agenda, and there were some who reflected reluctance to this extracurricular activity. But that hesitancy was quickly replaced by excited anticipation when we arrived at a beautiful, but war-torn, old historic castle in Beaufort.

Mr. Bartimes was joined by approximately 100 other citizens of the village in hosting the Sixth Armored people at a very lovely reception in the castle. Mr. Bartimes assembled the group, and from prepared notes started making a speech. He welcomed us to his town, thanked us for liberating his little home country from the Nazis 40 years ago. He praised the American soldiers who risked their lives, who withstood the dreadful winter of 1944 away from their own homes, etc.

Soon, he was speaking from the heart, never once referring to his notes. He said: "Speaking in behalf of the Association of Luxembourg

Buchenwald Prisoners, I am so happy to have this opportunity to thank you for our lives. Each year on the 11th of April we lay a wreath on General Patton's grave, commemorating the day your division entered Buchenwald. As long as we live, we won't forget what you did for us, we owe you our lives and freedom." As tears streamed down his face, he explained that he was a prisoner of Buchenwald for two long years, and that there were now 35 survivors of Buchenwald in Luxembourg.

We were proud Americans that day. We knew that there were at least 35 Europeans who were, indeed, thankful that the Americans had interceded against the Nazi domination.

Not a single soul complained about the "extracurricular" trip that day.

Chapter Fifteen
RETURNING TO THE FRONT

I had put all my worldly belongings in a duffel bag and was headed for someplace in eastern France. Most of my war bounty, like foreign money, pistols and Iron Cross medals, was gone when my personal affects reached me at Bar-le-Duc. At the first repo depot, I found that I was headed back to my old outfit in the 6th Armored Division. We traveled in large $2^1/_2$-ton Army trucks.

The Allied Forces had crossed the Rhine. The armored units were moving fast now, bypassing resistance and taking thousands of prisoners. Could it be true, I wondered, that we would get the Nazis to surrender before I got back to my unit?

It was close to the month of May and I was grouped with those assigned to the 3rd Army. We were shipped out the next day to another depot, I believe it was still in France. The next day we took a two-day ride across the Rhine, two-thirds of the way through Germany. We were in two $2^1/_2$-ton Army trucks escorted by Jeeps and a weapons carrier. We were not armed, but the MP escort was heavily armed. The long ride was not bad since we were not crowded. Our duffel bags in the center of the truck allowed us to take turns lying down and sleeping during the long journey.

We arrived at daybreak at division headquarters. We were given breakfast and sent to our battalions in the back of a weapons carrier. I arrived at the 50th Armored Infantry Battalion headquarters that morning. The other GIs were not from my company. The driver was trying to line us up rides to our own companies when I heard someone holler, "Andy, Andy Giambroni!" It was Silver, now a major. He had seen my name on the replacement list. "Come with me," he said, "bring your duffel bag and I'll take you to Company B." Major Silver then introduced me to officers of all ranks, including a colonel. They invited me into a large well-furnished room. In this room was a bar, with several kinds of European liquors. They were talking about the negotiations of the Allied surrender of the Nazi Regime. It was May 4, 1945. I believe the official signing was May 11, 1945. I think I saw tears in Silver's eyes.

Shortly thereafter, Silver and I drove about five or six miles to Rocklitz, Germany, where Company B headquarters was located. There I met Captain Haber, who had returned to duty. He had been the company's captain who was wounded in the taking of the Brittany Peninsula in July and August of the previous year. Haber was a good-looking man. He told me he had played football for Rutgers University before he went into the Army. We all had a victory highball, including 1st Sergeant Sanning. This was the last time I saw Major Silver. I've always been sorry I never kept up with him and many of the old-timers of Company B.

Very few of our squad were left; in fact most of the guys I knew when I first joined Company B in Nancy, France, were gone. Those left were happy to see me, though; we had a real welcome with drinks and food of all kinds. We had nice rooms in the quarters at Rocklitz. There was a stack of German weapons over six feet high with a 24-hour guard outside our CP.

The next 10 days we saw a steady stream of German soldiers on the road, heading west. They did not want to surrender to the Russians. They dragged along in their light blue or gray uniforms and soft gray caps. Some were bandaged from wounds they had received. This sight will live forever in the memories of the Allies and the Germans who witnessed this end of a war to end all wars.

I understand our famous General Patton was really hurt by being kept from entering Berlin first. In the following 50 years, such politics and foreign policy would repeat itself again and again. Patton was a certified prima donna, but he was a true warrior in every sense of the word. He knew his trade and will always be known as one of our great military generals.

The Russians were just a few kilometers across the Elba River, approximately 100 miles south of Berlin. Our Company sat there for another two weeks. The Russian soldiers came across the river and visited us on a couple of occasions. The German civilians were deathly afraid of the Russians and did not want them to occupy their city. "No gut ruski" was all through eastern Germany. We were forbidden to talk to Germans. If you were caught fraternizing you could be fined $65. I was never fined, however. I was never caught.

We took almost two weeks to pull the whole division back to the Frankfurt area. We bivouacked and lived in pup tents and we traveled on the autobahn as much as possible. It was springtime. Blossoms were on the trees and it was a great time to be alive. Naturally our thoughts of home and the good old US were stronger.

Many of the greatest heroes of the war are buried in any one of the many US military cemeteries in Europe. Their names were hard to remember, but the faces of these brave soldiers I'll never forget. The young man whose chest I pulled a big hunk of shrapnel out of rejoined our Company at Frankfurt. He looked me up to thank me and showed me the large scar on his chest.

While we were in the Frankfurt area, our Company was called to help with a train wreck. One train had run into the rear of another pulling box cars loaded with DPs from Poland and other Slavic countries. Seeing so many people killed or badly injured was still hard on my eyes, even after the battles I had been through. There were children, young babies, women and men of all ages. The train had crushed and pushed several box cars off the track.

I had gained most of my weight back and was in good physical condition. Our Company was preparing for our final formation. The 50th Armored Infantry was to be disbanded. Those soldiers with 85 points would be on their way home. I had 79 points and was slated for the second group to leave, whenever that was. Most of the rest of us were assigned to the 3rd Armored Division.

Our final 50th AIB formation was very memorable to me. Major General Robert Grow of the 6th Armored Division was to give the final inspection and his farewell talk. We all got haircuts. I even had my picture taken. The day finally came in late June. With all the spit and shine, we looked like soldiers once again. At the formation, a lieutenant and myself were escorted in front of the whole battalion and General Grow pinned the Silver Star medal on the lieutenant and me. These medals were for gallantry and bravery in combat. He stepped back and saluted both of us. He then shook our hands and had a big smile on his face. I really believe he recognized me from when we met at the Our River. I once read that General Bradley had said something like this:

"Bravery is when you are scared shitless and still able to perform what you think is right." So many GIs fell into this category. It was a great honor to be singled out of a bunch of great American soldiers.

Within 24 hours we all had our orders. Most of our company was headed for the 3rd Armored Division stationed just south of Frankfurt. The 6th Armored was disassembling many of its battalions. The 50th Armored Infantry Battalion now lives only in history books. Most of the men I knew in Company B were sent in many directions. I know one of the cooks of our company had been in the Army several years and was one of the first to go home. I gave him a 3-objective microscope I had picked up in Zeitz, Germany, and also gave him forty dollars in occupational money for his trouble to take it to California. He lived in Fresno, which was about 300 miles south of my home. Unfortunately, the beautiful microscope was not to be had. When I returned from the war and caught up with him, the cook from Company B told me that the port authority had confiscated the instrument. He was an alcoholic of the worst kind so I don't know if that story was true.

To this day I have not seen any of the soldiers I really knew in Company B of the 50th Infantry Battalion again. After 53 years I found out that the 6th Armored Division has an association. I contacted the executive secretary, Ed Reed, and immediately joined in hopes that I would meet someone I knew. So far I haven't met any of the old squad but I did meet approximately 300 great guys of the Super 6th Armored Division. We all have gray hair and our stomachs are not as flat as when we were spearheading for that man Patton. Oh what memories would I love to share if I met someone from the squad, Major Silver or 1st Sergeant Sanning or such. I still have hopes. We are the old guys now, the senior citizens.

In a little over a month, I was transferred from my old battalion company with no one I really knew. I was getting homesick every time I thought of transferring to another unit. Playing softball and reading animal husbandry books helped to slow the anticipation of going home. I had a new address but was assured that the mail would catch up with me. That's one thing I can say about the Army—they did everything for you, including forwarding mail to wherever you were stationed.

The extra winter clothing and wool blankets we had were turned in to the quartermaster. Travel light, we were told, you will be issued a new uniform at your next station. We had heard about the new Eisenhower jackets that fit snug at your waist. Yes, I thought, I would like to have one of those.

Before I left I heard that they had found Lejoy's body in the woods of Arloncourt. He had been missing in action since the battle in January. I could only hope that the little guy with the mangled legs had gone on to live. That battle at Arloncourt cost many brave American soldiers. Today, Arloncourt is a small farming village in eastern Belgium. The history of the battles fought there hopefully will never be repeated. The long, sloping fields that lead into this quaint village are etched in my memory. How quickly your attitude changes when you no longer have to dodge artillery shells, when you don't have to look and listen when you step out of a doorway or watch your buddy's back for him to watch yours. In some ways wars stop like turning off the key on an engine.

A couple of "foxhole generals" (Vargas and me)
talking over the war's end while returning to Frankfurt
from the Elba River where we had met the Russians.

Sergeant Edward Gerberry, squad leader of the 1st Platoon,
Company B, 50th Armored Infantry. Wounded in Belgium,
Sergeant Gerberry returned to the unit towards the end of the war.

Sergeant James Holsey, squad leader of the 1st Platoon,
Company B, 50th Armored Infantry. Also wounded in Belgium,
Sergeant Holsey was returned to the unit towards the end of the war.

Chapter Sixteen
THE OCCUPATIONAL ARMY

Traveling through Frankfurt to our new home with battalion headquarters of the armored infantry of the 3rd Armored Division, I was at amazed at how badly the Allied Forces had destroyed the city. I don't think I saw many intact buildings. The allied Air Force had made crumbles of one of Germany's largest cities. Traveling through town you could see where the bulldozers had made a path through the rubble where streets once were. With all this damage to Germany, the Allied ground forces had had difficulty in defeating the Nazi Regime.

The German population was now facing something worse than war—starvation. People in the cities especially had little or no food. Those in the agricultural areas had stored some food: potatoes, beets, turnips, apples and some cereal grains. Older people and young babies could not stand malnutrition. German funerals were more present in the immediate postwar months.

The war was still going on in the Pacific and our rations were cut back. We received our share of powdered milk, powdered eggs, powdered potatoes and other backup rations. The GIs did not throw it away—I can still see them scraping the powdered foods of their mess kits into the little buckets of young (7 to 12-year-old) German children. These little kids were already showing signs of malnutrition—little potbellies, sunken eyes and sallow skin. The GIs would go back to their billets, which were German homes still in living condition, and French-fry potatoes and apples in the grease they got from the mess sergeant. Several of us stationed in a little farming dorf a short distance south of Frankfurt would trade K rations for three or four chicken eggs. This little village (Rot-Am-See) accepted the GIs. They were glad to hear that they would be occupied by the Americans, not the Russians.

I was assigned to the AIB headquarters as an operations sergeant. My MOS was 814 but I would not get the rating or pay until I made up my mind to sign up for a tour of duty. The Lt. Colonel, commander of this armed infantry battalion of the 3rd Armored Division, knew I was high in points and would have an opportunity to go home in a few months. He brought into the battalion a young buck sergeant and told me to teach him the operations of an armored infantry battalion. He said, "You can change your mind." I was bent for hell on my path to become an animal doctor, though, and would get all the books I could on raising farm animals from the literature sent into our battalion.

Each morning we would fall out and have roll call, duties for the day and breakfast. I ate a lot of pancakes and we had our share of little children come to share the powdered eggs or mush from our mess kits. We would do some calisthenics and the athletic officer would often organize a softball game or 50–60-yard dashes. All of the latter was not mandatory. We had guard duty and some of the crew would patrol in Jeeps and even half-tracks. Curfew was about 9:30 in the evening. We were allowed to go to the beer garden but not allowed to sit with or speak to the Germans. This was the nonfraternization regulation that was still in order. I drank the flat German beer long enough until I liked it. Even the red, dark beer became one of my favorites. The German people were hard-working individuals. They impressed us by doing their farm work with one ox and one horse to pull the farming equipment, such as it was.

It was well into July by now and I took every trip offered to me. One trip I took was when about six other GIs, me and two officers went to Bavaria. This is a very beautiful part of Germany. A lot of Bavaria borders Switzerland and, like Switzerland, has large Alps. We went to Berchtesgaden, Hitler's summer home, and we visited the little town where all the townspeople put on the passion theater. This is the life story of Jesus Christ. Every person in town plays a part in the production and it is put on every ten years. This trip was for only three days. The next trip was to Paris, but it would have to wait.

The 3rd Armored Division had been given the presidential citation for action with the 1st Army during the war. Our battalion would stand as honor guard in Frankfurt for the new President of the United States, Harry S. Truman. Eisenhower and Truman walked a foot or so in front of me as our battalion representatives stood at attention. Harry Truman had a big smile on his face as he walked through the long line of GIs.

He was on his way to a meeting with the Allied Forces to divide Germany. The division had been already done at Potsdam by Roosevelt, Churchill and Stalin. Truman and Reagan will always be my favorite presidents. They both spoke their own mind.

Around this time I received a letter from my hometown barber. He had not heard from his parents in Frankfurt since 1939. He had made several attempts to contact them and then the war had shut him out. He asked me in his letter if I would look them up and enclosed a letter to them written in German. He also gave me the address of his father's home and barber shop in Frankfurt. In a few days I had some business in Frankfurt and asked the Colonel if I could take his Jeep and driver after we finished our reporting in Frankfurt to find these people. He said, "Don't get lost and get home by dark." We, the Jeep driver and myself, found the older couple in the rubble streets of Frankfurt. Their home was still intact, but the nearby buildings had been hit by Allied bombs. The couple were in their late 70s, the father weak and bedridden. The mother was taking care of her husband as best she could under the circumstances. They had a neighbor who spoke English and did the translating for us. She told us food and nourishment had been very scarce. On that I said I would be right back. We toured Frankfurt and found an artillery outfit stationed there. We went to the kitchen, found the mess sergeant, and told him we needed some food. He said he would have to have his officer okay it. When the officer asked what outfit I was with, I said the 3rd Armored. That was the magic word. I was only in the 3rd Armored a short time, but this division had some strong clout.

We were lucky to find the home again, everything was so torn up. We brought them some flour, powdered eggs, potatoes and powdered milk. The mess sergeant even threw in a small can of real coffee. The mother was so happy she tried to give me a diamond ring. I said, "No, you might have to trade that for food or medicine." I gave her the letter from her son. They both read it and cried. I told them I would try to get up here again, but this was not to happen—I was getting ready to go with the next group. Before I left this elderly German couple, the mother insisted I take her large topaz ring and a small bundle of letters she had written to her son through the years. I mailed the letters through the GI post office to her son in California. When I finally got home, one of my first visitors was my friend the barber. I showed him the topaz ring and he recognized it immediately as his mother's ring. I told him his mother wanted him to have it. I understand that a few years after that he had gone to Germany but his mother had passed away. He did get a chance to visit with his father. I've never heard from or seen him since. My life took me one way and he went another.

August was just marking time when we heard of the atomic bomb dropping on Japan. I believe VJ day was on September 2, 1945. It wouldn't be long before a lot of us would be going home. I did get that trip to Paris, for five days. We had one train day to Paris, three days in Paree and a one-day train trip back to Germany. Paris is a beautiful city. I have been back there twice in recent years. There is so much to see and enjoy in this historical city. I took that trip to Paris, however, through the courtesy of the US Army.

We were all decked out in our new Eisenhower jackets with a special green stripe on the shoulder. This green stripe was on the uniform of all GIs who were from the combat field. This insignia was familiar to the French civilians and most of all to GI MPs (military police). The French showed their gratitude over and over at the outdoor cabarets. We could hardly buy a drink. The first night in Paris was wine, women and song. The MPs helped us home—I guess we had celebrated too much. The following evening we happened to run into one of the MPs who had brought us home the night before. He said I was dancing at a local bar, but too many people were stepping on my hands. The next day I thought they'd also stepped on my head.

The French government gave us the very best hotels. The hotels the GIs stayed in I cannot afford today when I visit Paris. The French people had suffered under the Nazi occupation. French women who lived or slept with German soldiers had their hair shaved off completely. This was before wigs were popular. There were Frenchmen who had collaborated with the Nazis. They were turned over to the courts and sent to prison. There also were some who sold themselves for money or greed. They were never heard from again.

The French are a proud race of people. Their country has been used for a battleground more in this century than they wanted to endure. The Almighty must have intervened and saved the great city of Paris. The countryside of France still shows some of her battle scars. Paris, with all its artistic and historic wonders, has been spared the

September 1945—Me and two buddies of the Occupational Army near Frankfurt hold the famous M1 Gerand rifle.

Holding the M1, a superior rifle of World War II.

damage of war. You know, 53 years later, the German countryside has many great autobahns, underground electrical power and telephone communications. The tall modern buildings in their cities and the automobiles are top of the line. The last time I was in Europe, it was hard to believe that Germany had lost the war and these other European countries had been on the winning side.

We had no problem with the German people in our area. We were the Occupation Army. They just went about their business. In our area it was agriculture. They had small numbers of livestock, including cattle, swine and some sheep. Their small fields, 20–30 acres, were planted with hay, vegetables and maybe a small orchard. I would say this was the average size farm in Germany at the time. If a farmer had a hundred or 500 acres, usually it was him and one or two other families operating the farm. They worked, daylight to dusk, every day except Sunday when they went to church. The Lutheran Church seemed to be the church of choice in our area south of Frankfurt. There were very few tractors, since fuel was in a shortage. The farming equipment and wagons were pulled with horses, oxen or one horse and semintel milk cow. They used many different combinations for dray animals. The people of the agricultural areas did have food stored, milk from the cows and eggs from the chickens. The Germans of the metropolitan areas were starving. The small amount of food they got was from relatives or friends in the agricultural areas. The Germans love coffee, but there was no coffee. They would roast cereal grains, wheat or barley, and grind it for hot coffee. It tasted terrible, believe me.

The orders waiting for me when I returned from Paris were the orders I had been waiting for—I was going to a staging area in preparation to go back to the States. The staging area was in a small city, Heidenheim, just a short distance from where I was.

Chapter Seventeen
GOING HOME

The city of Heidenheim had not suffered as much damage as other German cities. Heidenheim was more of an agricultural town. The people here either owned or worked in the rolling, fertile soil of the area. Some of the homes on the edge of the city were billets for the GIs. Many of the GIs I met there had been overseas for three or four years. Points to go home were given for time overseas, battle stars and decorations received overseas.

You can imagine the thrill of knowing you were going home. It was October and we played a lot of touch football and softball. The only formation was roll call in the morning. They did take the big Army trucks on sightseeing tours of the area. We were about halfway between Frankfurt and Munich. The area was part of the western section of Germany that the United States would occupy. The Red Cross and other organizations were furnishing writing materials and lists of library books available. I got some good books on animal husbandry and spent a lot of time reading and writing home and to the Colorado State University Vet School.

I took a few of the sightseeing tours. That time of year the cattle and other livestock were still out of the barns and in the fields. During the late fall and winter months the livestock were kept in the barns. These barns were in direct association to the home (living quarters). This was true in most parts of Europe, and the feed for the animals also was stored in these barns or in stacks, adjacent to the barns. The German people did not waste anything. Even the corners of the fields had been tilled with hand forks. Women were commonly seen spading up every inch of tillable soil. I saw breeds of cattle in this part of the country that were not introduced into the United States until after the war. Their breeds were used for many purposes: food, milk and as oxen to pull equipment.

There was lot to be learned from these people whom we had just conquered. They were above average intelligence; how they could have let a guy like Hitler put them into such a disgraceful situation is hard to guess. Hitler and his cohorts had brought Germany out of the depression using gang-like tactics. Human lives were expendable if they could take their wealth, whatever it might be. It was not only Jews he preyed on. Hitler was after royalty or anyone considered to be wealthy. He did increase Germany's economy, through illegal means or otherwise. He grabbed the lion's share of the wealth, from his country and neighboring countries. The biggest portion of wealth was to build his military strength. Baiting the German people with the hopes of good times, he organized the deadly Nazi party.

We were still under the nonfraternization order, but the troops going home did visit with more of the German civilians. The younger Germans I could not trust—they still had the combat look in their eyes. My mind was on getting home and getting on with my life. Both day and night I said prayers to the Almighty and my dad, thanking them for getting me through this ordeal. I remember once I had asked my dad if, when I got my education, he would help buy a cattle ranch. He got up out of his chair and walked over to a window and said, "Look out there. It's a whole wide world. Get your share of it and you can buy your own cattle ranch." I never forgot those words. First I had to get an education. The next steps would follow, if I kept my nose pointed the right way.

November was upon us and the cold, damp weather had moved in. It was just a year before that I had been fighting the same Germans not far away in the Lorraine Country of France. Fighting around Metz I really discovered how brutal war was. It isn't fancy flags, trumpeting horns, the quickness of a snare drum, or sharp-looking uniforms on parade. The brutality of war is the picture that no artist can paint. I still believe in a strong military for defense, not aggression. Like my high school coach once said, "We use this play when everything else fails."

We heard in the chow line that orders were in and we would be traveling to France on the railroad. They didn't tell us it would be in boxcars—the old 40-and-eight boxcars used in World War I. Forty-and-eight stood for 40 hommes (men) and eight chevaux (horses). Forty men crowded into these boxcars, I always wondered where in the hell they put the horses. We traveled day and night for three days through

October 1945—Sharing a bottle of wine
on our way home from Heidenheim, Germany.

October 1945—Going home from Heidenheim, Germany.
Nonfraternization with German people was still being enforced.

southern France sleeping on our barracks bags. You know, we didn't care, we were going home. We lived on K rations from Uncle Sam, French bread, cheese and wine given to us from the crowds of French people who would meet the train in the towns and small villages as we passed through. Marseilles, a French city on the coast of the Mediterranean Sea, was our destination. Here we were to board a ship to take us to the New York City harbor.

We were in Marseilles for two or three days; there we were given a physical exam and any vaccination shots due and put into groups of 200–300 GIs. We were paid the full amount of our overseas pay at that time. It was nice handling US currency again. I put it into my money belt and told myself that is where it would stay until I got home. Approximately 5,000 troops climbed aboard the USS Hermitage, a captured Italian ocean liner. We loaded onto the ship up the gangplank and down into the second deck below. The accommodations were a lot better than the English ship I had come over on. We all had a bunk to sleep in, even those who were spaced two to three bunks high. The bottom bunk, I remember, was almost on the deck.

The ship was run by our own Navy and the food was first-class. We had beef, chicken and even turkey on Thanksgiving. The voyage was 11–12 days and we spent Thanksgiving on board eating a great meal with all the trimmings, including ice cream for dessert. It was a long time since I had had such a great meal. I had two bowls of ice cream. The Navy cooks enjoyed feeding us dogfaces and they had a lot of mouths to feed. They did it in shifts according to the deck you were on. Their system was organized and everyone got plenty to eat.

I was very lucky, I never did get seasick on the trip home. The young man in a bunk across from me wasn't that fortunate. He had a special puke bucket set up next to his bunk. The medic gave him some pills, but he threw those up in short order. I made a big mistake after dinner one day. I told him the pork chops were out of this world. That one statement started a series of dry heaves. I felt really bad about that. I don't think anyone can get any sicker than seasick. By the time we reached New York, he was feeling better and about 20 pounds lighter.

Coming into the New York harbor we passed close enough to see the Statue of Liberty. I'll bet all 5,000 guys had a lump in their throat; I know it was a great sight to my eyes. I saw many guys with tears running down their faces. My eyes were pretty wet, too, but you know, infantry sergeants don't cry. I know it sounds like a lot of bull, but at that moment in my life I really realized how lucky I was to be an American citizen. I also knew the big debt we owed for the lives and mangled bodies of those Americans who have fought the battles of our country.

It was the first day of December 1945, a day all 5,000 GIs aboard will always remember. A tug pulled the ship into unloading ports. The Navy had several deck groups go back to their quarters because the ship was listing to one side from all the GIs on the top waving and whistling at the crowd. There was a band playing and hundreds of people were waving American flags. The noise was loud as we filed down the gangplank with our duffel bags. We walked single file a short ways and were loaded onto a passenger train. It was mid-morning when the train went over a bridge to the neighboring state of New Jersey. The train took us to Camp Kilmer, close to the city of New Brunswick, New Jersey. There we had a late lunch and were assigned barracks at the camp.

October 1945—Can't tell you why
I have a feather in my cap.

Chapter Eighteen
THE LAST LEG HOME IN THE US

The first evening home in the good old US couldn't be spent in an Army camp, but no passes were being issued at this mustering-out camp. Someone told us a guy in our barracks knew where there was a hole in the chain-link fence. I found this guy with a handlebar mustache who said he and two other guys were going to town at dark. I took a shower, put on my snappy Eisenhower jacket, and was ready for an evening in New Brunswick. I met with the three guys and we were through the fence like corn through a goose.

It was no problem for a GI to hitchhike a ride on the first of December 1945. We visited several bars and danced with any lady who would dance with us. Later that evening we decided to have our picture taken together. With two guys standing and two guys sitting, the picture was put on an 8 × 4 card. I was standing with another guy about 6′ in height, and the guy with the handlebar mustache and a paratrooper were sitting on chairs. Why I tell you of this photo now, is that going on a Sonoma County horseman ride some 20 years later, I found out the guy with the handlebars, tailor-made Eisenhower jacket and flowing mop of hair was the same guy I had been riding horses with for several years. His wife had the picture of the four of us, the same photo I had sent to my mother in California. In 1945, I remember, each one of us gave five dollars for his own copy. As of this writing, Brock, his wife and my mother all have passed away, but those pictures still remain.

I'm sure Brock was on the C-47 Army plane that took off about December 7, 1945, from the Newark airport. These planes were not jets, they were prop workhorse transport planes from the US Air Force. We sat in bucket seats along the wall of the fuselage. There were about 25 of us headed for Sacramento, California. One captain was in charge of us 24 enlisted men. Our first stop would be in Knoxville, Tennessee. In Knoxville we stopped for fuel and got out to go to the latrine and

December 1945—The picture worth a thousand words.
Rear left standing, Andy Giambroni. Front right sitting, Brock.
Who knew after 20-some years these two men would become
friends and later discover they had their picture taken together
in New Brunswick after spending a night on the town.

stretch our legs. In a short while we were up and flying again, only to circle back to Knoxville to have one of our engines checked. This time we were on the ground for about an hour. The pilot said just before take-off, "Don't worry, these C-47s won't ever quit you." That evening we landed in Dallas, Texas. Our gang spent approximately four hours in the coffee shop while mechanics worked over our flying machine. We then were off for Sacramento, with a short stop in Tucson, Arizona, for fuel and a latrine break.

When we arrived at the Sacramento Airport it was just 24 hours after we had left Newark, New Jersey. While waiting for the bus to take us to Camp Beal, Marysville (approximately 75 miles north of Sacramento) the Captain came up to me and said, "Sergeant, are you sure you're entitled to all the decorations on your blouse?" The GIs called these ribbons fruit salad. I said, "Sir, you have all of our service records in that big folder, why don't you check it out yourself." I never heard from him again.

The Army checked me medically more ways when I was discharged than when they drafted me almost three years before. A first sergeant with hash marks half-way up his sleeve gave me a resounding pep talk, but he knew I was headed back to college to get a degree as doctor of veterinary medicine. Throughout the days I had to report to several formats. Each evening I would hitchhike about 70 miles south to Woodland and spend the night with my brother who was a doctor of vet medicine. We talked about me getting back to Colorado State University and starting my studies again with the hope of getting accepted to the next class of vet medicine in the fall of 1946.

On December 11, 1945, in a small chapel at Camp Beal Army base in Marysville, about 50 Army veterans received their honorable discharges. Afterwards I couldn't get on the highway fast enough to hitchhike a ride to Woodland and my brother. I had just the few clothes the Army had given us and 300 dollars mustering-out pay. The Army was history to me, like turning a page in a book. I was to do more thinking for myself from then on.

A person on his own runs into plenty of ups and downs when turned loose on the whole wide world. This was surely one of the times in my life for decisions. These decisions had to be honest, not necessarily easy, but made with my full effort and attention. It was fun going on calls with my brother, delivering calves and foals, getting milk cows up with milk fever, and giving dogs and cats distemper vaccinations. I remember my brother suturing up the badly cut front leg of a horse. This did remind me of the torn bodies of the young infantry soldiers. He deadened the legs of the horse with Novocain and sutured him standing up. My brother was more than an average veterinarian and I still hear veterinarians brag about his diagnostic skills with animals. His favorite animals were horses and he spent many hours attending to all breeds.

Joe and I went to Oakland to my mother's home for Christmas. There I met with my younger brother, Frank, two older sisters and their husbands. I was home only 11 days before I was on a train to Colorado. I was to ring in the New Year, 1946, in the snow-covered city of Fort Collins, a beautiful town about 65 miles north of Denver at the foot of the Rocky Mountains.

On January 2, I entered into the GI Bill of higher learning. I had saved $1,500 from my Army career, money that would come in handy. I went to the office of Dr. Newsome, the dean of the school of veterinary medicine. He said he had received several letters from me when I was in Europe and informed me that they would only be accepting 60 students in the freshman class for the coming fall. He made it clear that the school would be partial to students from Colorado and neighboring states first. He said that California was a rich state and should have its own school but that they would be taking a few California students because there were only four schools of veterinary medicine west of the Mississippi.

Dr. Newsome pointed to a wall full of files in his large office and said, "I already have had 1,500 applications to the school. Many of these applicants do not have the prescribed courses to qualify for acceptance." I said, "Dr. Newsome, you may have 1,500 people that would like to be accepted to that class, but you don't have one that wants to be in that class any more than I do." He had a big smile on his face and said, "I'll write down some of the courses you should take to complete your qualification." He told me that I had three quarters, including the summer, to complete these courses. "How well you handle these courses," he said, "will be a factor in getting a chance at the fall class." My hand was shaking when I took the list of classes. I was really excited—I was finally going to have a chance at the profession I had dreamt of.

I was 22 years old and ready to dedicate the majority of my time to studying, taking notes and spending time with students who had taken these courses. I studied until midnight and one in the morning on many evenings. I had to develop methods of studying, memorizing and retyping my notes. I had to do these things because I knew I wasn't a natural student. I thought that I had developed unique methods of studying until I saw several books that included my techniques and some that I wish I had used in my college days. My goal was A's, nothing else. I needed these grades if I had a chance to make the fall class, especially being from California. I ended the three quarters with all A's and two B's and made the freshman class in the fall of 1946. I graduated in June 1950 with a degree of DVM (Doctor of Veterinary Medicine).

Surviving the brutality of war and making my own dreams come true proved to me that if you try hard enough and sometimes long enough there is nothing a person cannot accomplish. I've told young people, including my own kids, that you never know your own limits until you challenge them.

History of Company "B," Ninth Armored Infantry Battalion,
Sixth Armored Division
by Col. Jim Moncrief and Other Members of Company "B"
Close Combat or "Eye-Ball to Eye-Ball" Contact with the Enemy
Reprinted by permission of J. Moncrief

It is known that certain soldiers have tougher jobs than others. The guy who is a typist or clerk in Theater Headquarters certainly has a far easier life than the ammunition carrier in a mortar squad of an Infantry outfit. The fellow standing guard at the entrance into the Corps Headquarters bivouac is not subjected to the rigors of combat as is the rifleman of an Infantry squad. The experience of the truck driver moving supplies from the rear echelon to the Field Army "dumps" can't be compared to that of the tank driver, who suddenly finds his tank face to face with an enemy tank as he makes a turn at a road intersection. The radio operator at the G-3 Tent of a Division Headquarters enjoys comforts and conveniences never experienced by the machine gunner on a half-track. The cook in an Infantry Battalion Headquarters does not face the enemy's rifle and machine gun fire as does the rifleman of an Infantry squad.

Yes, the job of being a soldier who faces the enemy eye ball to eye ball should be singled out as being the toughest, the most dangerous, with greater risks of life, the least comfortable, and requiring the most physical endurance than any assignment in the Army. He may be an Infantry rifleman, who, while facing enemy fire, advances with his fellow platoon members to knock out a machine gun at the top of the hill. He could be a member of a tank crew as his tank gets in a more advantageous position to fire at an enemy tank emerging from the woods. He could be an engineer racing onto a bridge to cut the wires leading to an explosive which the enemy, in his retreat, had installed to detonate the bridge. He may be an Artillery Forward Observer, positioned in the Church Steeple, who is, by radio, bringing fire from the heavy guns to bear on an enemy advancing within fifty yards of the church, and the area occupied by friendly troops.

In addition to the extremely hazardous conditions of enemy fire faced by the Close Combat Soldier, he is constantly subjected to personal hardships, far more serious than those experienced by other soldiers. He is forced to live with: hazardous weather conditions, cold food, little sleep, no bath for days, maybe weeks, little chance for a change of clothing. Because of the elements of Mother Nature he endures: the snow, the ice, the rain, and ankle-deep mud, while in clothing wet to the skin. Much of the time he operates from the cramped and limited space of a narrow, hastily constructed foxhole, which he has dug, sometimes into the frozen ground, but always with a small inadequate hand shovel. Frequently, the foxhole is dug while he is under fire from the enemy.

The casualty level of these types of soldiers is much higher than among others. In most units where Close Combat was a daily routine, the casualty rate during WWII was over 100%. Being a replacement assigned to such a unit engaged in Close Combat has to be one of the worst experiences to which a human being can be exposed. Nothing can produce a feeling more awkward, insecure, humbling, and sensitive; and at the same time generate so little gratification, appreciation or understanding. Therefore, the Replacements who join a unit where "eye-ball to eye-ball" soldiering is routine deserves double credit or recognition.

For a nation to be victorious at war, those soldiers in close combat with the enemy must be successful. The ultimate objective of a nation at war is to impose its will upon the enemy. To do that requires that nation's forces to conquer the military forces of the opposing nation. Such cannot be accomplished without routing the enemy forces from his position on the ground. The same ground must be physically occupied by the ground (close combat) troops of the enforcing nation. The entire effort of a nation involved in a war is centered around producing and delivering the where-with-all—equipment, material, including trained manpower for close combat duty—required for those same soldiers to achieve their objectives.

The various activities of the Command and Staff functions of each of the seven or eight echelon levels (from Company, Battalion, Regiment, Division, Corps, Army, Army Group, to Theater) of command in a Theater of Operations are centered and focused on insuring that the soldiers in close combat with the enemy accomplish their mission. Organizing the soldiers into battle-ready units, assigning appropriate missions, establishing areas of operation, coordinating support fire from the air and sea as well as that from ground artillery, arranging for logistical support to include medical, food, communication, transportation, ammunition, and maintenance are but a few of the duties of these various echelons of Command. It is important to note that all of these Headquarters are in existence solely for the purpose of supporting the soldier who is in close combat with the enemy.

An oversimplified analogy of the above may be described as the relationship between a golfer and his caddie. The golfer is responsible for withstanding the pressure, the weather, the hazards, and other problems, while at the same time, making the all-important shot, the end result, on which his success or failure depends. Just as the combat soldier is responsible for the end result, so is the golfer responsible for the final score. The caddie is responsible for toting the golfer's bag, advising the golfer concerning club selection, determining distances, holding his umbrella in case of rain, drying the club grip, raking the sand traps—for the total support of the golfer. In the Army scenario, there are thousands of caddies, or people in support roles, for each close combat soldier. Note the one significant, yet ironic, difference in the comparison. In the world of sport, the golfer is the one who, not only makes the score, but also he receives the prize money, honors, and the press reviews accompanying his victory. In the Army, the close combat soldier while he achieves the victory, gets very little financial return, seldom is recognized or honored by the Congress or the press nearly so much as some of the "caddies" who have furnished him his support.

In today's Army, the Combat Infantryman's Badge is awarded to the Infantry soldier. This award is given across the board to all members of an Infantry Regiment or Battalion. Not every soldier in an Infantry Regiment or Battalion is subjected to the "eye-ball to eye-ball" combat described in the paragraphs above.

It is my opinion that the Army should create a new Decoration called the "Close Combat Medal" to award soldiers, not restricted to Infantrymen, who perform their duties under conditions of "eye-ball" contact with the enemy. Further, I believe the Replacement Soldier who earns the Close Combat Medal should receive additional special recognition.

Biographical Sketch

JAMES S. MONCRIEF JR., COLONEL (RETIRED), USA

Born: June 7, 1912, Manchester, GA
Graduated High School, 1928, Albany, GA
Graduated University of Georgia, 1933
Commissioned 2nd Lieutenant Infantry Reserve, June 1933
Active Duty, Civilian Conservation Corps (CCC):
  Six Months, 1933–34, as 2nd Lt.
  One Year, 1937–38, as 1st Lt.
Educational Advisor (as civilian) in Civilian Conservation Corps,
  1934–36
One Year Active Duty in Regular Army under Thomason Act,
  1936–37 with 22nd Infantry Regiment, at Ft. McClellan, AL
Employed as Field Investigator and Interviewer in Legal Department,
  the Coca-Cola Company, Atlanta, GA, 1938–41
Active Duty, 2nd Armored Division, Ft. Benning, GA, February
  1941–42
Cadred to 6th Armored Division upon activation, February 1942
Company Commander and Regiment Adjutant, 50th Armored
  Infantry Regiment
Assigned (as Capt.) as G-1 (Personnel) of Division in 1943–
  promoted to Major and Lt. Colonel; served in that capacity as
  Lt. Colonel throughout the War in Europe
Chief of Staff of 6th Armored Division upon deactivation in
  September 1945
War Department Manpower Board, Atlanta, GA and the Pentagon,
  1946–47
Integrated into Regular Army, 1946
Student, Graduate School, University of North Carolina, 1947–48

Army Secretary, Committee on Human Resources, Research and
Development Board, Department of Defense, Pentagon, 1949–52
G-4, Third Army Corps, Ft. McArthur, CA, 1952-53
Staff Negotiator, United Nations Command Military Armistice
Commission (UNMAC) Panmunjom, Korea, 1953–54
Asst. J-1, Far East Command, Tokyo, Japan, 1954–56
Professor, Military Science and Tactics (PMST), West Virginia
University, Morgantown, W. Va., 1956–59
Military Attached to Nicaragua, 1960–63, during Somoza's regime
Professor, Military Science and Tactics (PMST), University of
Wisconsin, Madison, 1963–65
Retired as Colonel, July 1965
Executive Vice President, Chamber of Commerce, Monroe, NC,
1966–68
Vice President, American Bank and Trust Co. (later United Carolina
Bank), Monroe, NC, 1968–77

In the infantry, after a tactical field exercise, we had what was called a critique to summarize the performance of the exercise just performed. We talked about the good things we did as well as the mistakes we made. We learned a great deal from both the positive and negative results. A few weeks of severe combat can be compared to a lifetime or more. I hope you can visualize, from this book, the reality of war.

Fifty years after the European war, I took a trip to Europe with my wife, daughter-in-law and two sons. This was the best of a life that has passed too quickly. I am 77-years-old, living in a body that should have been turned into a scrap yard. The golden years are made up of memories of the years we filled to the fullest. Wars have been going on for centuries and I can see that the defensive action of war is a necessity to those defending their own people. We must stop there, however, and not act like policemen protecting other people or countries, at the expense of our own citizens' lives. If it is a political or religious problem that cause these atrocities, I can't see us being much help. This type of domestic battle will simply continue as soon as NATO leaves.

Wars have left an indelible mark on the millions of men and women who have served in the United States military this past century. It did not matter whether you served in combat or not, it still made changes in your life. Some of the changes were for the better, but most were negative for the young men and women whose minds and outlook on life were abruptly changed in a matter of days, when war was declared. The American people have always supported the people in the military. They have given the youth and joy of their lives to the causes of our great nation. When we saw firsthand the dead and mangled bodies of American GIs and the enemy, the question in the minds of thousands was—why?

Everyone who fought in the Bulge was a hero in his own right. I would like to single out in this critique a few people, products and equipment that stood out in the European victory: The striking power of our P-47s in low altitude bombing of enemy positions when visibility and weather were in our favor. The 90mm direct fire cannons that replaced the 75mm on our tanks and TDs (tank destroyers). This weapon made a big difference when introduced into the Battle of the Bulge. The coordination and the accuracy of our artillery battalions was greater than words can describe. Our own artillery was a giant morale builder after we had taken several hours of enemy punishing artillery fire. The M1 rifle, 30-caliber semi automatic. This outstanding rifle, also known as a Gerand, had firepower and accuracy up to 500 yards. The GI boot grease that kept the dampness from soaking your boots in the cold Ardennes. Frozen hands and feet were one of our big casualty lists–keeping these appendages dry and warm was a constant job. The K ration that a soldier could survive on when our mess sergeant couldn't get hot chow to us. This small ration box was the size of a small book, waterproof and durable. An infantry soldier would carry two or three inside his wool-lined jacket. The fabric of wool was one of the heroes of the European War, especially in the sub-zero weather of Belgium. The many garments and blankets made of wool were a survival component of the United States military. Besides the wool blankets, there were overcoats, shirts, pants, caps, gloves and scarves, all using wool. Last, but not least, is the single combat unit, the infantry soldier. His performance against outstanding odds and an enemy with more experience and training was amazing. He surprised the enemy and our own commanding officers, with his ability to survive.

I must add to this critique how well the Germans have done in the unbelievable reconstruction of their country. The autobahns, underground electrical services and towering modern buildings will surprise all who have seen war-torn Germany. This is not as evident in France, Belgium, Austria or Italy. More of the scars of war can still be seen in these countries. The economy in West Germany also seems to be better than their neighboring countries. The equipment, vehicles and merchandise displays have the appearance of a strong economy.

The German people will not volunteer anything about the war; however, they will answer questions politely and to the point. With further questioning, they will give you some of their opinions. I have found the German people to be very intelligent and industrious. They are a very proud and honest people in the business I have done with them in recent years. It seems odd that they are the same people that General Patton told us to destroy by any means or way, without getting destroyed ourselves. I am sure those who fought the Orientals in the Pacific, Korea and Vietnam, are reminded of these wars when we buy and trade products with these countries.

Time—days, months, years and a new generation—smoothes out the past, encouraging us to be a closer world. In all of our hopes and prayers, a world of peace and the pursuit of happiness will be available to all.

The soft long notes of taps by the bugler brings memories of both our military and civilian heroes. They remind us of what life is all about. They seem to ask us not to let their deaths be in vain. Many of the heroes of war never had the luxury of a full life. I can still see many of their faces even though I didn't even know a lot of their names. Many of them died instantly to artillery shells or direct cannon fire. There were those who died painful deaths, with shrapnel and bullet wounds in their torso and abdominal cavities, and those who lost their extremities from freezing, which in turn, led to their deaths. The anti-personnel mines took their toll in wounds to the extremities and death as well. The treacherous teller mines destroyed vehicles, tanks and the occupants who manned them. Sometimes it seems there were more ways to die, than to live.

The Battle of the Bulge was only five to six weeks in length. There are not very many veterans alive today who fought through the span of the Bulge. Twenty thousand American soldiers died in this battle. This is the largest number of soldiers killed in any battle fought by the US Army, other than the Civil War. More than 77 thousand casualties were racked up during the Bulge. The enemy, however, suffered many more casualties and deaths than the (mongrel-bred) GI. Hitler underestimated the mixed nationalities of the American troops. We survived overwhelming odds and weather conditions and still were aggressive. My son asked me, after seeing a movie of D-Day, "Dad, were you in any battle that horrible?" I said, "Son, I was with my B Company in three attacks of that caliber. Two at Mageret, Belgium and one at Arloncourt, Belgium. We were knee deep in the snow, charging into all types of deadly weapons. The Battle of the Bulge or anything like it, I pray our military does not have to face again."

I remember returning from the hospital to my company, deep into East Germany, where I saw the massive destruction of roads, bridges, railroads and buildings done by the Allied Air Forces and the retreating

German Army. Large cities like Frankfurt were in crumbles. Bulldozers pushed paths through the large cities and towns to allow military vehicles to pass through. German civilians were in critical need of medical attention and medical supplies. Availability of food was a number one concern for the German civilians. In the last weeks and months of this terrible war, the very young and very old civilians suffered the most.

Traveling to Eastern Germany, we bypassed long lines of German prisoners of war. They were walking, guarded by GI MPs (military police). The MPs were riding in jeeps and weapon carriers and were well armed with pistols, rifles and machine guns. There didn't seem to be any fight left in these prisoners. The once very powerful German war machine had been defeated even though it wasn't official yet. Most of the prisoners had the soft gray/blue caps on, dirty uniforms from the dust on the road. Still, a large percentage of them were clean-shaven but needed a hair cut two or three months ago. Their hob nail boots clattered on the pavement of the autobahn. Some of them were real young and others had gray hair starting to show. They were of all sizes and stature, not many having the luxury of a covering of fat around their mid-section. We observed a group of prisoners at a break—it seems everyone in the German army smoked cigarettes. All of them still had their belts on with a canteen and mess tools. There was no smiling or laughing, just some quiet conversation.

The big question still goes through the minds of many people—how can so many intelligent people end up in a situation like this? So many of them wanted to surrender to the American Zone. The Krauts were extremely afraid of the Russians and how they would be treated in their processing. I have been told that the Russians did not release many of their war prisoners for four or five years. They were used as slave labor in Russian prison camps. This was told to me by German families that did not get their fathers, sons, brothers or husbands from the Russians for this long of a period of time. I know the old saying goes, everything is fair in love and war, anyone can start a war, it takes at least two people to perform love. Fifty years has smoothed over the deep and hard feelings of that war. It seems that we keep finding ways to get into other wars. There are people who know what war really is and that people will have to suffer and will do everything, in every way, to avoid the real meaning of war.

I visited the US Army Cemetery in Luxembourg more than 50 years after the war. Standing beside the memorial monument, I looked out over the 55 acres of white crosses and Stars of David. This is just one of the many United States Cemeteries in Europe. There are many for our Allies and those for our enemy. What a price we all have paid to gain freedom and peace. At the head of this 55-acre cemetery is buried the great warrior himself, General George Patton. He fits in well with other heroes of ours. This is the same general who had no time for death. He was like a football coach: "Hit 'em hard, hit 'em quick, stop only to fill the gas tank." He waited until he ran the enemy off the field of battle before he would die.

My family was with me at the Luxembourg cemetary that day. They knew that each cross and Star of David stood for one of my heroes. I stopped to read many of their names and the names of their outfits. Very few could I remember, but I could see many of their faces when I closed my eyes. A great number of these heroes are buried all over Europe. They will stay there to remind all who travel through that they paid the toll fee. When we were ready to leave, I stopped at the top of the steps and turned toward the cemetery. I don't know who did it, but when I looked out at all the crosses, Stars of David and Gen. Patton's grave just below me, the sound of Taps rang out from a recording above in the monument. My throat and chest tightened and my eyes filled with flooding tears. I tightened my teeth together—I don't believe in grown men crying. My family closed against me and grabbed my arm and hands. My heroes I would say good-bye to, and promise them that I'll always speak out on how horrible war really is.

ACKNOWLEDGMENTS

Many thanks to the people who contributed their time and talents to the creation of this book:

Jennifer, my daughter-in-law, who did much of the editing and made many of my notes readable: thank you, Jen, for helping me get this 50-year-old story told. My son Jess, the artist, who wouldn't settle for anything less than perfection. It's nice to have a son like him. Anyone writing a book for the first time had better have a wife like Bev. She encouraged me always so generations to come would know the horrors of war. Joe, my other son who has a natural gift for art, was an enthusiastic supporter of this project. Mike Benny, a great friend to Jess and awesome illustrator: thank you, Mike, for building the picture on the book's cover. Excerpts from Colonel Jim Moncrief's two books add greatly to this story of an infantry soldier and war itself. Thank you, Colonel, for the permission to use your experience. Ralph Emerson Hibbs, MD, author of "Tell MacArthur to Wait." He served in Bataan and told of the atrocities of the GIs held in Japanese prison camps. Al Meglio and members of the F Company, 386th Infantry of the 97th Division. I trained with these soldiers in California. They, too, saw combat in the European theater. The editor and staff at the Veterans of Foreign War magazine, Richard Kolb, Robert Widener and Pat Brown. These people went out of their way to help me in my research of infantry soldiers and the 6th Armored Division. Walt Peacock, a close friend for 50+ years and purveyor of volumes of books on World War II. Ralph Byers, infantry soldier of Company B 50th Armored Infantry Battalion 6th Armored Division. Ralph was severely wounded in the upper leg and pelvis near Metz, France. Eva Krois, who assisted with typing and working evenings to help pull this together. Mark Wilson, a good friend, student of infantry weapons and infantry soldier of the National Guard. Will Williams, retired 1st Sergeant of the 50th Infantry Battalion. He is a true supporter of the infantry soldier. Paul Giambroni supplied considerable knowledge on the printing of the book. I would like to thank Tom Moore Photography for the picture on page 20.

1945—Sergeant Andy Giambroni, Paris, France.
A retired veterinarian, the author now resides
in Red Bluff, California.